Reimagining Environmental History

Reimagining Environmental History

Ecological Memory in the Wake of Landscape Change

CHRISTIAN KNOELLER

UNIVERSITY OF NEVADA PRESS *Reno & Las Vegas*

University of Nevada Press | Reno, Nevada 89557 USA
www.unpress.nevada.edu

Library of Congress Cataloging-in-Publication Data
Names: Knoeller, Christian, 1954– author.
Title: Reimagining environmental history : ecological memory in the wake of landscape change / by Christian Knoeller.
Description: Reno : University of Nevada Press, 2017. | Includes bibliographical references and index.
Identifiers: LCCN 2017012485 (print) | LCCN 2017029807 (e-book) | ISBN 978-1-943859-51-1 (paperback) | ISBN 978-1-943859-52-8 (cloth) | ISBN 978-0-87417-604-9 (e-book)
Subjects: LCSH: American literature—Middle West—History and criticism. | Ecocriticism. | American literature—20th century—History and criticism. | Ecology in art. | Ecology in literature. | Conservation of natural resources in literature. | BISAC: Nature / Environmental Conservation & Protection.
Classification: LCC PS273 (e-book) | LCC PS273 .K56 2017 (print) | DDC 810.9/977—dc23
LC record available at https://lccn.loc.gov/2017012485

The paper used in this book meets the requirements of American National Standard for Information Sciences—Permanence of Paper for Printed Library Materials, ANSI/NISO Z39.48-1992 (R2002).

First Printing

Manufactured in the United States of America

To Julie, my wife and partner, who day by day
and year by year has whole-heartedly engaged
in the many ways we have shaped this book together.

If nature writing is a vital conversation through time,
one that teaches us about the land and the ways in which
it has been imagined, discovered, settled, forsaken, and redeemed,
the subject of that conversation is the intersection between
nature and culture in a particular place.

—MICHAEL BRANCH AND DANIEL PHILIPPON, *The Height of the Mountains*

Leaving natural breaths, sounds of rain and winds, calls as of birds
and animals in the woods, syllabled to us for names,

Okonee, Koosa, Ottawa, Monongahela, Sauk, Natchez,
Chattahoochee, Kaqueta, Oronoco,

Wabash, Miami, Saginaw, Chippewa, Oshkosh, Walla-Walla,

Leaving such to the States they melt, they depart,
charging the water and the land with names.

—WALT WHITMAN, "Starting from Paumanok"

Contents

Preface

We try to save what is passing, if only by describing it, telling it, knowing all the time we can't do any of these things.

—W. S. MERWIN, "W. S. Merwin, The Art of Poetry"

Reimagining Environmental History presents an ecocritical study examining representations of landscape change, from nineteenth-century artist-naturalists to pioneering ecologists to twentieth-century poets, Native American novelists, and contemporary literary naturalists associated with the Midwest. While the scope is expansive in terms of period and genre, the volume as a whole is unified by focusing on the depiction of environmental history in the region. Many of the authors addressed cross-conventional disciplinary and cultural boundaries: whether they are artists writing about wildlife, Native American novelists tracing the connection of indigenous languages to nature in particular places, ecologists writing literary natural history, poets expressing environmental sensibilities, or contemporary literary naturalists incorporating anthropology, archaeology, geology, and linguistics into essays about nature and place. While each chapter explores the work of a single author in depth, cohesion of the whole is maintained by making connections between and among these writers, as well as by considering how their work speaks to current concerns such as stemming extinctions and perpetuating species biodiversity. Each of the authors examined engages place in direct and personal ways, informed by a sense of environmental history, whether in their own bioregion or as they sojourn through less familiar terrain, as well as when contemplating the *idea* of what is ancient, indigenous, and ongoing. For as Scott Russell Sanders observes, "To become intimate with your home region, to know the territory as well as you can, to understand your life as woven into the local lives does not prevent you from recognizing and honoring the diversity of other places, cultures, ways."[1] It seems high time to take stock of how this written record of environmental history might in turn inform our efforts toward preservation and restoration going forward—especially in an era of cascading environmental decline, when the very ideals of wilderness and "wild" are being redefined.

By examining landscape change depicted in several genres of texts spanning two centuries, this study is situated on the cusp of ecocriticism and environmental history. What unexpected insights might arise by analysis conjoining the two in light of repeated calls over the past decade for examining how these fields might interact in fruitful new ways (e.g., Cohen, Slovic, Bergthaller)? In what sense might they even prove to be mutually illuminating? Specifically, what insights could these two potentially synergistic disciplines provide together when addressing such complex and perplexing questions as how historical and cultural context shape our perception of, interpretation of, and response to centuries of landscape change? I hope that this work contributes to the ongoing exploration of territory at the confluence of two such complementary bodies of thought. Above all, what forms might ecological restoration best take today—and to accomplish just what enduring aims? Lest we underestimate the remarkable resilience of this living earth, consider the truth of naturalist Paul Gruchow's declaration that "one of the fundamental qualities of life is that it is organized to endure."[2]

What has made the representation of landscape change in the American Midwest especially prescient and compelling has been the severity and pace of the change. During the nineteenth and twentieth centuries the American Midwest became a crucible for forward-looking environmental thought, summoning the courage to hope that we might—indeed, we must—find a way to stay the forces of perpetual development before it is too late to save any remaining semblance of the endemic prairie biome and its legacy of fertile soils that have endowed the nation and the world with such agricultural abundance. As prairie and wetland biomes were converted to cropland, the pace of landscape change and its cumulative ecological impact became too rampant and widespread to escape notice. Technology assuredly played a central role in accelerating the process, as steam dredges for the first time enabled the draining of wetlands, steel ploughs broke prairie sod irreparably, and railroads transported crops to burgeoning urban markets from coast to coast. Taken together, such technological advances precipitated the region-wide *erasure* of entire wild landscapes with their self-perpetuating communities of indigenous plant and animal species.

All too often, however, when the erasure of native ecosystems across North America has taken place incrementally over many years, characterized by increasingly fragmented habitat and the gradual extirpation of wildlife populations, each successive generation perceives nature in terms of the diminished landscape that it has inherited (referred to as *shifting baselines*). We cross a daunting threshold when what has been lost is also forgotten, so that we lose track of the extent of that loss. The converse might be thought of as *ecological memory*: the interpretation of environmental history by a variety of writers representing ongoing landscape change in places they have lived or traveled in light of concerns about its overall extent and cumulative consequences. In this respect, earlier accounts such as John James Audubon's (1785–1851) from the first half of the nineteenth century provide an essential ecological baseline by describing a level of natural abundance almost unimaginable today. Many writers in the centuries since have lamented both the pace and extent of pervasive landscape change they personally witnessed in the region, while often attempting to reconcile their own accounts with the written record of past abundance. Their testimony inspires this reconsideration of writing about nature in the Midwest from an ecocritical perspective. The authors examined in this study each witnessed natural landscapes fundamentally disrupted or altered entirely, places that inspired them aesthetically, scientifically, and spiritually. In effect, each becomes a case study in ways of engaging and representing nature, envisioning environmental history, and conceptualizing stewardship.

The purpose of this book is to contribute to the scholarship of environmental history in several ways: (1) to "recover" the work of several of the region's critically overlooked literary naturalists, including Gene Stratton-Porter (1863–1924) and Paul Errington (1902–62), as well as to "discover" several more recent writers such as Paul Gruchow (1947–2004) and Elizabeth Dodd (b. 1962) who are yet to receive adequate recognition for their contributions to environmental history of the Midwest; (2) to explore connections and lineage among the region's literary naturalists from an ecocritical perspective; and (3) to reassess how the legacy of Native American history and culture has been represented in midwestern literature as well as how the region's indigenous languages evolved in relation to nature and are therefore inextricably linked to place.

Gene Stratton-Porter was an early and arguably seminal figure unfairly dismissed by literary critics of her time, yet she was also a groundbreaking, albeit self-taught, naturalist. Her fiction and nonfiction carry a prescient environmental message so compelling as to command our attention today. Moreover, the unprecedented restoration of her own land in Indiana, involving the transplanting of tens of thousands of native plant specimens, parallels Aldo Leopold's celebrated experiment, though it actually preceded his by a generation and arguably even foreshadowed recent developments in ecological restoration. Similarly, pioneering ecologist Paul Errington—a close associate of Leopold—has also been largely overlooked. Both recognized interrelationships between the members of ecological communities, appreciated the complex interdependence among species, and questioned the conventional wisdom about predation in ways that anticipated current thinking about the essential role played by keystone species, and the reintroduction of major predators such as wolves to portions of their historical range.

Important correspondences emerge when several authors write about visiting the same places and attempt to interpret ancient traces inscribed on the landscape by indigenous peoples: the pictographs in the Boundary Waters region of northern Minnesota and southern Canada, for example. In *Books and Islands in Ojibwe Country*, Louise Erdrich describes images painted on rock faces encountered while traveling through the area by canoe. Scott Russell Sanders and Paul Gruchow also write about their experiences contemplating the pictographs while canoeing there. Similarly, both Gruchow and Elizabeth Dodd have written about visiting the Medicine Wheel monument high in the Bighorn Mountains of Wyoming and reflecting on its enigmatic history, recognizing the difficulty of interpreting the significance of such sites fully since cultural memory had been disrupted—including lost or forgotten narratives of their creation, meaning, and function. Poets expressing environmental sensibilities such as Theodore Roethke (1908–63) and William Stafford (1914–93) contemplate the surviving cultural legacy of Native American cultures, as do contemporary literary naturalists such as Sanders and Dodd. I read Dodd in relation to celebrated essayist Loren Eiseley, while devoting the culminating chapter to Gruchow. Above all, I wish to explore their visions—individually and collectively—for preservation, restoration, and stewardship in the wake of the region's history of agricultural development, habitat loss, and environmental decline.

Consider the ecological legacy inherited by those born and reared in the Midwest today. Over the years I have invited university students to describe a place that had awakened their own appreciation of nature, and whose destruction by development they later witnessed; they reflect on the emotional impact back then and as they recount it years later. Reading their responses—most of them were born in the Midwest during the closing decade of the twentieth century—I am stunned by the litany of loss they describe. Psychologically leaning schools of environmental history recognize the emotional intensity of lamenting such a loss as deeply *personal* grief—as if we had lost a part of ourselves. Clearly we can no longer afford to dismiss as quaint or nostalgic accounts of environmental decline across North America, particularly those from the second half of the nineteenth century forward. Since the forces of climate change we face today continue to accelerate exponentially the rate of wildlife extirpation and species extinction, the alarm sounded by writers over the last century and a half takes on a heightened significance; indeed, their message has never seemed more timely or urgent.

> *In all our attempts to renew or correct ourselves,*
> *to shake off despair and have hope, our starting place is always*
> *and only our own experience. We can begin (and we must always*
> *be beginning) only where our history has brought us.*
>
> —Wendell Berry, "Ignorance"

On a personal note, I would like to describe what moved me to write this book. My wife and I reside in a Victorian farmhouse in central Indiana on the banks of Deer Creek, a tributary of the Tippecanoe and Wabash Rivers. Our house was built soon after the Civil War of rough-hewn timber frame posts and beams cut locally, probably from the property's own woodlot, as was customary in the days before commercial lumber became available. Floors are tulip poplar and red oak, wainscoting black walnut. Initials carved into barn doors by previous generations of residents date back to the nineteenth century. In its day, this homestead was in a small way the embodiment of Jefferson's vision of self-sufficient yeomen livelihood: a family farmstead surrounded by pastures for livestock and woodlots that provided ample natural habitat. By the time we acquired the property in the summer of 1999, adjoining acres had been cleared, ploughed, and planted with row crops for over a century, cultivated now in the manner of industrial agriculture with its massive machinery and petrochemical fertilizers as well as a suite of toxic herbicides and insecticides, in tandem with genetically modified seed. One of the central narratives of this book concerns all that has been lost in the process—as well as the implications for the forms of restoration that might take place today.

A third of a mile south of our house stood a small marsh—a circular depression toward the eastern edge of the upper field, not more than thirty feet higher than the banks of Deer Creek that Hoosier poet James Whitcomb Riley rhapsodized in his sentimental nineteenth-century lyrical verse. In fact, the real estate listing shamelessly yet aptly proclaimed that this was just the kind of place he would have felt right at home. That tiny wetland was an ecological wonder: a tangle of cattails the roots of which alone stood head high when water subsided each summer. Such a marsh represents a microcosm of environmental and evolutionary history: an island of deep time providing habitat uniquely suited to a suite of ancient species. In spring, snapping turtles that had burrowed beneath to hibernate emerged, carrying up to a foot of mud caked on top of their shells. Every summer night tree frogs, known regionally as peepers, raised a chorus that carried a mile or more. Deer bedded there, especially during hunting season, since the tangle of cattails provided nearly impenetrable cover. Most awe inspiring of all was that moment every year when huge dragonflies would circle upward by the hundreds in columns of tightly synchronized flight.

News reached me that the neighboring farmer had other plans for the small wetland adjoining our property. So when the day came that the cattails would be burned, I had been forewarned. It took just a few men to ignite the flames and spread enough accelerant to keep them burning the entire day. As I watched in disbelief, I was overcome not so much by the vile black smoke rising but by a sense of grief that gripped me so deeply I could hardly breathe. The process that followed was workmanlike, involving a backhoe and front-end loader, as the embers of roots reaching back millennia were buried in shallow trenches, and a system of drainage pipes or "tiles" was installed to funnel groundwater down to the lower field toward the river. Immediately, the peepers in their sonorous multitudes were silenced. The snapping turtles also disappeared that season. Memories of those swirling columns of dragonflies that once rose triumphantly above our home—emissaries from deep time, 300 million years ago—still haunt me.

Given the environmental desecration that I have witnessed personally, the preservation of wetlands strikes an especially deep chord. The burning and burial of the marsh seemed to me an unspeakable crime against nature that represents in microcosm the whole region's environmental history over the last century and a half. What cut to the quick was the utter erasure of a place that had awakened such wonder. This, I think, is a universal story, albeit localized, a wrenching reminder of environmental decline writ large. If a phrase like "declensionist narrative" evokes a whiff of cliché—a story we have already heard all too often before, or worse yet, nostalgia, that hopeless yearning for what is perceived to be irretrievably lost—so be it. Perhaps this is the crux of the problem. The moment that we relegate laments for the seemingly never-ending assaults on the environment to the category of "old news," and see calls for measures to stay or reverse such forces as an exercise in futility, is the moment we run the risk of losing hope itself.

Here on Deer Creek, the spring after the marsh was burned, we discovered, as unlikely as it might seem, tentative shoots of cattails rising from the remnants of roots that had been bulldozed and buried. This minor miracle (which, we would learn, would happen just once) was the working of what landscape restoration has termed *ecological memory*, the biological processes of regeneration enabling indigenous species to return and reassert themselves following disruption. We set about transplanting these remnant cattails onto our own land beside a spring below the house—first dozens, then hundreds—and on a soggy slope beyond the barn where none had grown before, at least not in our time, where they have since naturalized and a small colony of leopard frogs now thrives.

My fascination with natural history was undoubtedly instilled by formative, childhood experiences of exploring the outdoors. My earliest memories of engaging nature, however, are hardly an agrarian idyll, the place being Long Island, New York, not far from Levittown, the community that became the prototype for suburban development across the country immediately following World War II.

At about ten years of age, I began collecting butterflies that haunted feral edges of our neighborhood—islands of habitat preserved purely by chance, such as narrow meadows of native wildflowers thriving beyond the cyclone fence of the elementary schoolyard. To this day, I cannot imagine summers there without calling their erratic flight to mind. Among my prized specimens at the time were the Black and Tiger Swallowtails, Buckeyes and Question Marks, Mourning Cloaks and Red Admirals. Monarchs meandered in mid-September as autumn approached before beginning their long flight south. They appeared to be a consummate survivor—enduring an epic migration, over the course of two generations, to wintering grounds in Mexico. Given their longevity, in fact, these delicate insects could live all winter indoors and be released again when weather warmed in the spring. Who would have imagined that the seemingly resilient population of a species once numbering in the billions could utterly collapse only half a century later to below thirty-five million due to rampant habitat destruction coupled with the ubiquitous use of broad-spectrum herbicides? Indeed, the future of the species that had always ushered in our autumns is presently imperiled by agricultural practices that have eradicated incidental habitat along the edges of cropland across the continent, erasing a once vast network of microhabitats, especially the milkweed essential to sustaining such monumental migrations.

Among the magnificent moths that captured my imagination as a child, the hauntingly beautiful luna occasionally made what seemed to me spectral visitations, all the more precious by their scarcity—maybe half a dozen personal sightings in my entire lifetime. I even became intimately acquainted with the seemingly miraculous metamorphosis of the spectacular cecropia moth, a species that enchanted Hoosier naturalist Gene Stratton-Porter above all others. Having found a huge cocoon tethered to twigs in brush along the road, I brought it home and kept it in a large jar, hammering holes in the lid since even creatures that hibernate whole seasons, I reasoned, need to breathe, then set it on the nightstand beside my bed. All winter, sleeping side by side, we changed from night to night, its embryonic wings layered in quarters too tight to unfold. I still remember watching it hatch: a slow crawling toward light, the pained extrication, that tentative unfolding, spreading out, attempting balance, finding patience for pigments to dry and the intricate architecture of veins to stiffen.

Even while still a sixth grader, I marveled at traces of ancient abundance that unaccountably remained in the heart of suburbia. Walking the bank of a shallow creek linking a pair of modest, man-made lakes on Long Island one summer, I found the entire channel choked with juvenile American or "glass" eels, anticipating upcoming spawning runs in subsequent years, before their drastic decline beginning in the mid-1980s. That experience echoed a still earlier one: a forest floor teeming with bright orange salamanders (*Notophthalmus viridescens* in its intermediate, red eft terrestrial stage)—inordinate hordes of Eastern newts swarming along woodland creeks, presumably seeking to breed along the shore of Lake George in upstate New York.

I again glimpsed traces of age-old abundance as an adolescent, after our family moved to a wooded neighborhood in northern New Jersey. On sweltering summer nights, sleeping on a screened-in porch, I would drift off listening to the chorus of untold numbers of katydids, crickets, and other nocturnal insects, trying to imagine the dimensions of a wild universe that surrounds suburban existence even now with its primordial sound. Waking before sunrise, sometimes I would wander the woods where our dead-end street began, following an intermittent stream to the little bog that then still held developers at bay. One morning, I encountered bevies of box turtles on the move, dozens on dozens marching across the forest floor to lay eggs, as they have for 300 million years, before returning to burrows deep enough to wait out the next winter.

But perhaps most arresting of all, in the midst of suburban sprawl on the edge of a small historical park just twenty miles from home, a remnant population of native brook trout still thrived. I had been shown the way there by a high school friend on the condition that I never reveal the location to anyone. But I brought my father there once, and later a fiancé, that they too might witness what seemed such a timeless wonder. This population of indigenous fish—so tiny yet fierce—had apparently survived since the last Ice Age in the little woodland brook, bearing to this day their exquisite markings in hues rivaling the intensity of jewels.

To put into perspective why such an unintended refuge for native brook trout seemed so utterly unlikely and precarious, consider the state of neighboring waterways in New Jersey at that time. In the mid-1960s, the Passaic River that flowed through my hometown had been named one of the ten most polluted in the nation. If you waded to the waist against its sluggish current to cast for catfish, carp, and the occasional stray pickerel, the water was so opaque that you could not see as deep as your own knees. And as implausible as it may seem today, decades after the passage of the Clean Water Act in 1977, a toxic cocktail of chemicals was continuously dumped into such waterways. Possibly the most egregious affront of all were the large blocks of a dense Styrofoam-like substance, presumably an industrial by-product of some kind, recklessly discharged and routinely left to float downstream. Suburban families living along the banks would scavenge pieces nearly the circumference of compact cars to use as primitive rafts, or even to chisel into dugout canoes. I would question such a recollection, if it were not so utterly unforgettable—and the memory still so vivid to this day of that dull cinnabar flotsam shot through with a hint of autumnal orange. How could this happen, I wondered, in such unthinkably close proximity to that seemingly pristine trout stream? Struggling to make sense of such incongruities ignited my lifelong penchant for reading about natural history and ecological topics, particularly environmental history.

Several seminal works by literary naturalists have fundamentally reshaped my understanding of ongoing landscape change. I will describe briefly how each catalyzed my interest in environmental history, as well as touch on their continued relevance. Such books take on heightened significance today in light of ecological quandaries accelerated by climate change. But first, to establish a baseline for gauging the cumulative impact of ongoing environmental decline, consider the fabled bounty of the New World testified to by European explorers. My initial encounter with this written historical record came while I was still in high school. *America as Seen by Its First Explorers* by historian John Bakeless purported to provide a composite portrait of North America in what was then presumed to be a primal and largely unadulterated state, one deemed "unspoiled" and "untamed." The book opens with an earnest attempt at addressing the history of indigenous peoples in North America; being published in 1950, however, the work predates—and consequently failed to anticipate—a postcolonial perspective.

The veracity of Bakewell's expansive narrative is dependent on the credibility of those who observed America's early natural wonders firsthand from the remarkable abundance of verdant forests teeming with wildlife to watersheds incredibly rich in anadromous as well as freshwater fish. Accordingly, *America* incorporates a host of primary source documents, drawing extensively on previously unpublished archival materials such as period letters and journals. Organized chronologically, the volume proceeds from ships' logs kept while mapping the Atlantic coastline of the Eastern Seaboard, to the quixotic overland journeys of conquistadors, including Cortés and Coronado, and later expeditions into the interior by river from Champlain at the dawn of the seventeenth century to George Catlin on the upper Missouri in the first half of the nineteenth. Analogies to Eden abound in these early accounts, as explorers attempted to depict what astonished them in terms that would not be dismissed as implausible by their contemporaries, let alone those of us reading in the wake of centuries of profound landscape change. The impression *America* left on me was the legacy of monumental ecological alteration on a continental scale—a logical precursor to the concept of ecological memory to be explored in the course of this book.

Reimagining environmental history involves recognizing cumulative changes to the landscape such as the destruction of natural habitat, whether the result of extractive activities such as logging or development for agricultural use. "Improvements" that typically accompanied initial settlement—whether felling trees, clearing fields, draining swamps, or breaking prairie sod—are shadowed by native species that would readily return in our wake, an instance of ecological memory in the biological sense. Elements of prior ecosystems can often be rejuvenated, at least partially, from traces that remain. Roots and seeds, dormant perhaps but still viable, awaiting rejuvenation or germination, are among the agents enabling regeneration. Corresponding processes of plant succession and species diversification ensue the moment any land is left fallow. In the parlance of modern landscape restoration, *ecological memory* refers to the dynamic biological capacity of ecosystems to recover from disruption.[3] Yet the concept of ecological memory can also be extended and reconstrued in cultural terms: our collective recognition of the environmental consequences and cumulative extent of landscape change.[4]

I can date my initial acquaintance with many of the concerns addressed in this book to the moment in the late 1970s when by sheer serendipity I purchased from a used bookseller in Eugene, Oregon, an early trade paperback edition of Peter Matthiessen's *Wildlife in America*. It seemed at first glance to be a standard volume of natural history, presumably celebrating the magnificence of the continent's megafauna, perhaps above all trophy game. Yet the book's benign title was belied by cover copy declaring it to be "the first history of man's effect on . . . extinct and vanishing species."[5] The contents proved to be profoundly troubling: a historical recounting of the precipitous decline in wildlife since European colonization of North America, as the edge of frontier settlement swept ever westward. This groundbreaking and now classic depiction of environmental history, originally published in 1959, struck me with the force of revelation.

Matthiessen's saga of environmental history and its litany of wildlife decline was mapped in my imagination onto a visceral impression of the topography of North America that I had acquired while bicycling coast to coast during the nation's bicentennial only a few years earlier. We pedaled for six months from the woodlands of Appalachia to the Pacific headlands along the Oregon Coast. In Kansas cafés, descendants of Great Plains pioneers likened our transcontinental ride on two wheels to that of their kin, whose prairie schooners had halted for them to settle the region more than a century before. Unwittingly, we had pedaled our way into the legacy of Manifest Destiny, or so it seemed to some of them. In truth, we were simply cyclists at the mercy of prevailing headwinds and afternoon temperatures soaring above one hundred degrees. A day riding across that prairie landscape might cover seventy-five miles or more of gently rolling slopes of grassland, cactus, and sunflowers, with an occasional antelope startled from grazing. In the course of the journey, I had in fact internalized majesty and desolation, as well as developed a fledgling sense of interlocking biomes spanning the Great Plains, followed by basin and range, and ultimately a succession of mountain ranges culminating with the Rockies and Cascades.

In retrospect, Matthiessen's *Wildlife in America* seems prescient in many respects, addressing issues that seem equally pressing today: extinction and the spread of invasive species, as well as the cumulative consequences of toxic herbicides that have become more pervasive in the decades since. Emphasizing regional suites of species, *Wildlife* is organized geographically, presenting a catalog of population decline, regional extirpation, and ultimately absolute extinction of several species. It contrasts the case of American bison, brought back from the brink, with the once bounteous passenger pigeon, which has been lost forever to extinction. The counterpoint of their respective fates, he reminds us, ranks among the founding narratives of the modern conservation movement.

The causes of wildlife decline that Matthiessen cites are legion: monumental habitat loss due to settlement, development, and widespread overgrazing, as well as ecological destruction wrought by extractive industries such as mining and logging. Moreover, nearly the entire prairie biome has been converted to cropland. The forces driving the environmental decline he described, of course, remain very much with us today. Consider the ongoing use of herbicides to eradicate large swaths of exotics. The current fervor for battling particularly aggressive species can unleash massive quantities of highly toxic substances into the environment—ironically in the name of conservation. The poisoning of ecological communities in order to return to some previous state—often idealized—ultimately proves unattainable. Eradicating invasive plants that have outcompeted native species can exacerbate environmental disruption, impacting entire ecosystems already impaired by development. Moreover, industrial agriculture has become so dependent on the use of proprietary herbicides in conjunction with genetically modified seeds that common weeds have already become resistant. Some ancient species such as horsetail have begun evolving into "super weeds" that appear to mutate and thrive on repeated applications of the very chemicals intended to control them. Other weed species will inevitably develop similar resistance. Should such trends continue, farmers will likely be forced to spray their crops with increasing frequency. This pattern of environmental degradation by agricultural chemicals follows logically from Matthiessen's dire warnings—as well as those raised three years *later* by Rachel Carson in *Silent Spring*—almost sixty years ago.

About the same time I first read *Wildlife in America*, Barry Lopez published *Of Wolves and Men*, the visionary work that secured his place among the preeminent American writers of our time concerned with the intersections of nature, culture, history, and ethics. *Wolves* is grounded in a keen appreciation for both cultural and environmental history. It contrasts Native American reverence for the wolf with European demonization that over many centuries has often led to its extermination. Environmental history implicitly provides a framework for interpreting how culture mediates our relationship with nature, as well as how changes to landscape and wildlife populations due to development have fundamentally altered ecological communities across the continent—including the Great Plains. "In North America, and elsewhere," he declares, "as human civilization affected the distribution and food habits of wolves—by killing buffalo, for example, and putting domestic cattle in their place on the ranges," many regional subspecies have been lost.[6]

Wolves, as Lopez explains, readily adapt their hunting strategies to pursue particular prey species as well as adjust to variations in terrain. Yet this versatility makes them increasingly subject to human ire and ever more vulnerable to retaliation by hunters and ranchers. Accordingly, reintroduction of major predators such as wolves to portions of their historic range remains highly controversial, especially when herds of domestic livestock such as cattle or sheep are at stake. While programs reimbursing losses sometimes succeed in diffusing tensions, there are still periodic calls to resume campaigns of extermination, especially when wolves are also perceived to diminish populations of game that hunters prize. When wolves prey on the same species as sportsmen—particularly deer, elk, moose, and caribou, as well as Dall sheep—"management" strategies as ruthless as aerial gunning have periodically been reinstated. Despite the concerted efforts of wildlife biologists, environmental advocates, and even tribal entities of various Native American nations, some sport hunters maintain that the wolf is public enemy number one, and go so far as to label its ongoing eradication an act of conservation. "In evolutionary terms," as Lopez notes, "wolves and men developed along similar lines as social hunters and were in competition for the same game."[7] In the context of deep time, today's hunters are expressing a very old grudge indeed.

Although terms such as *landscape change* and even *environmental history* had less currency when *Of Wolves and Men* was published in 1978, both concepts are implicitly central to its analysis of ecological memory. Discussing the presumed extinction of seven American subspecies of the wolf, for example, Lopez explains how the inbreeding of fragmented, remnant populations had effectively erased place-specific genetic distinctions, even as the wolf was being extirpated from much of its historic range. Concerns for genetic diversity take on increasing urgency today given each species' evolutionary potential for adapting to ongoing climate change. "A significant part of the genetic reservoir that once represented one of the more adaptive mammals on the face of the earth," he concludes, "is now gone."[8] Such capsule narratives of environmental history occur throughout the book.

This was also the era when Wendell Berry raised his clarion call for ecological conscience and ethical relations toward nature in *The Unsettling of America* (1977), a collection of polemic essays questioning the long-term efficacy of industrialized agriculture and stressing the environmental hazards of increasingly chemical-intensive methods of farming. By contrast, the traditional American family farm had long reflected Jefferson's ideals and entailed intimate acquaintance and continuous interaction with the natural world. The increasing consolidation of agricultural production in this country by corporate entities during the twentieth century, in Berry's estimation, represents a devil's bargain, by which the very fabric of agrarian life as well as attendant values such as stewardship were dismantled, and culminated in the patenting of genetically modified seeds in conjunction with proprietary chemicals including herbicides and highly toxic pesticides implicated in colony collapse disorder among bees as well as sharp declines in many avian species.[9] Environmental advocacy, he warned, that focuses primarily on setting aside remote tracts for preservation—no matter how high-minded—was liable to get mired elsewhere in the bureaucratic inertia of institutions that generally govern land use.

Berry contended that how we have related to nature in the past is an essential element of human history, since culture itself is fundamentally informed by how we respond to the places we inhabit and the landscapes that sustain us. "Our biological roots as well as our cultural roots," he declared, "are in nature."[10] Much like Thoreau, he posited that "wilderness" was "indispensable" as a "cultural model" and in a sense even represents a measure of civilization itself.[11] Moreover, *Unsettling* in a sense anticipated the ways in which "wilderness" and "wild" are being redefined in our time. Whereas Matthiessen and Lopez focused on the nexus of wildlife and cultural history, Berry widened the scope to encompass vast landscapes dedicated to agriculture, including those of the American Midwest. His central arguments remain relevant and compelling today, and similar sentiments can be seen to reemerge with equal vehemence in the essays of subsequent literary naturalists addressed in this volume, particularly Scott Russell Sanders and Paul Gruchow.

In addition to his essays, I savored Berry's poetry, as well as that of his contemporaries who also expressed a sense of environmental history, including two figures examined in this book—Theodore Roethke and William Stafford—as well A. R. Ammons and Gary Snyder. Having grown up among the monumental stumps left when loggers felled old-growth forests of the Pacific Northwest, Snyder, in addition to his widely admired poetry, penned early essays recounting deforestation and desertification in Asia, much as George Perkins Marsh had done a century before. I first studied Ammons's work in earnest as an undergraduate in the mid-1970s, actually making a pilgrimage to Ithaca to interview him at Cornell as part of a capstone project my senior year. As fate would have it, Archie was a dear friend of my father. Beginning in the late 1940s, the two exchanged poems and letters for decades. My intellectual debt to both of them is immense and especially evident in my discussion of Roethke.

In the decades since, more recent volumes of place-based literary criticism have continued to deepen my appreciation for environmental history, including *Surveying the Interior* by Rick Van Noy (2003), *Believing in Place* by Richard Francaviglia (2003), *Not Just Any Land* by John Price (2004), *The Midwestern Pastoral* by Bill Barillas (2006), and finally *Deep Map Country* by Susan Maher (2014). Each in its own way has helped blaze the trail I hope to follow here.

In *Surveying the Interior: Literary Cartographers and the Sense of Place* (2003), Rick Van Noy examines the perception of place in relation to the making of maps—graphic depictions of topography and space, subject to cultural assumptions about landscape and nature. What animates his analysis is the interplay between outward formulations of topography through cartography and inward experiences of place that can be expressed to some extent in various forms of literature. "The sublime," he writes, is "a way to express the wild (even eternal) space within the known (if soon to be altered) place."[12] By highlighting the element of time inherent in any accounting of environmental history, Van Noy signals the point of departure for the present study.

Van Noy recognizes, moreover, that literary representations of place entail a kind of "environmental reading," one that incorporates preexisting narratives—presumably including those of indigenous peoples.[13] The "sublime," he asserts, even as constituted in European intellectual and literary traditions such as Romanticism, "bears comparison with native stories that communicated 'spirit of place.'"[14] Accordingly, reading environmental history in North America, as well as writing it, necessitates a kind of cultural triangulation across time. Written accounts of landscape change on this continent since European contact, I would argue, are inherently situated in this intercultural matrix. Consequently, interpretations of environmental history conveyed by literary works carry an ethical valence when ecological disruption registers as a matter of conscience—what Van Noy terms "the responsibility that ensues when we project language onto nature, the ethical obligations generated by writing about place."[15]

Believing in Place: A Spiritual Geography of the Great Basin by Richard Francaviglia (2003) consists of accounts of personal journeys throughout the region interspersed with philosophical reflections as well as metaphysical conjecture. He considers how various indigenous peoples and European settlers have perceived and represented nature and place as sacred, as well as the role of language and story in perpetuating cultural perspectives connected with particular landscapes. The region's "human story," he asserts, "is largely scripted around the drama of North American/European American conflict…in addition to a rich Native American past."[16] The writers examined in the following pages, as we will see, often juxtapose these cultural paradigms, and at times seek to reconcile competing worldviews as powerful and potentially complementary ways of interpreting our relationship to the rest of the natural world.

The Midwestern Pastoral: Place and Landscape in Literature of the American Heartland by William Barillas (2006) is grounded in seminal studies such as *The Machine in the Garden* by Leo Marx (1964) that established how the pastoral ideal in both its sentimental and more sophisticated manifestations (as well as resistance to it) has been central to American literary tradition for centuries. Yet this study of regional literature is equally informed by the works of influential ecocritics, perhaps above all *The Environmental Imagination* by Lawrence Buell. Barillas brings an ecocritical perspective to bear on place-based literature while addressing the works of the region's novelists, such as Willa Cather and Jim Harrison, as well as poets such as Theodore Roethke and James Wright. He also acknowledges the significance of a few important new voices among midwestern naturalists, including Paul Gruchow, the subject of my concluding chapter. Barillas's study provides a precedent for my own analysis of cultural and literary representations of our relationship to nature in the region as depicted in several genres across time, touching frequently on themes essential to reading environmental history in North America, such as species biodiversity and extinction, succession and restoration, and the legacy of indigenous history and culture—the very constellation of topics to be explored in the coming pages.

Not Just Any Land: A Personal and Literary Journey into the American Grasslands by John Price (2004) melds memoir with interviews of major writers whose work has explored their own identities in relation to the prairie landscape. Price's account of reencountering the grasslands of his Iowa youth is a perceptive and intimate reflection on the ways relationship to nature and particular places has shaped not only his own identity but those of writers he admires, including Linda Hasselstrom, Dan O'Brien, and Mary Swander. Issues including ownership, appropriation, and "authenticity"—especially with reference to indigenous ancestry—precipitate serious soul searching by William Least Heat-Moon and Price alike. As such, *Not Just Any Land* sets a standard for how to address matters of such a vexing and culturally sensitive nature.

Similarly, the concepts of "ecopastoral" and "settler imaginary" posited by Tom Lynch (2015) place the wholesale transformation of prairie biomes in the broader context of European colonization and domination.[17] The eradication of buffalo and subsequent introduction of cattle has triggered degradation of much of the untilled short-grass prairie by widespread overgrazing; indeed, Lynch sees excessive ranching as an abomination of the European agrarian traditions rooted in Romanticism. The tragic result has included utter disruption of entire indigenous ecosystems, accompanied by pervasive environmental decline. The reintroduction of bison—especially in cooperation with today's Native American nations—offers the prospect of restoration, a theme taken up in several following chapters, particularly those devoted to William Stafford, Scott Russell Sanders, and Paul Gruchow.

Finally, *Deep Map Country: Literary Cartography of the Great Plains* by Susan Maher (2014) elevates Heat-Moon's *PrairyErth* to frame an examination of literary nonfiction of the region, including the writing of Loren Eiseley, Wes Jackson, John McPhee, and Wallace Stegner. Maher acknowledges Price's study, and her theoretical framework incorporates the concept of literary cartography explored by Van Noy in *Surveying the Interior*, in conjunction with the premise of the multifaceted *deep map* posited by Heat-Moon and developed more recently by leading ecocritics including Randall Roorda. Maher's interdisciplinary perspective proves particularly attentive to various dimensions of time, including not only the centuries of exploration and white settlement, but many preceding millennia of indigenous presence in the region, and considers its deeper geological, evolutionary, and environmental history. Maher's groundbreaking study arrives at a layered matrix of interpretation attuned to human cultural history closely associated with place as well as the region's ecological legacy—the same array of themes to be traced in the following pages.

Reimagining Environmental History is organized as five pairs of chapters, juxtaposing writers linked by historical period, literary genre, or theme. The first set, for instance, pairs artist-naturalists John James Audubon (1785–1851) and Gene Stratton-Porter (1863–1924). Chapter 1, "The Making of a Conservationist," begins by considering how two commonly held and seemingly incompatible interpretations of Audubon can arise from reading his legacy in terms of *today's* ecological imperatives. The first view holds that as a profligate hunter on the frontier, he forfeits any credibility with regard to environmental ethics. However, he chronicled changes to major game populations as ranges waned across much of North America during his lifetime, and was actually among the first to publicly acknowledge the prospect of species extinction. The countervailing position hails him as a prophet of conservation. In this view, his work represents the seminal contribution of a quintessential nineteenth-century naturalist. Underlying both views is our difficulty in imagining historic wildlife populations that so dramatically exceeded those encountered today. His writings provide an essential reminder by establishing a baseline for the abundance across much of North America during the first half of the nineteenth century.

Chapter 2, "Envisioning Restoration," traces the trajectory of Gene Stratton-Porter's career (1863–1924) as prolific author and pioneering nature photographer. Her reverence for nature led her to lament environmental decline and ultimately to advocate for conservation through the restoration of endangered native species. Extrapolating from environmental destruction she had personally observed, Stratton-Porter cautioned that unchecked development could have dire consequences, including the extirpation of wildlife locally and possibly the extinction of species entirely. Her enduring contribution constitutes an ode to an older order: a testament to what she perceived to be the sacred text of nature.

The second pair juxtaposes twentieth-century literary naturalists Paul Errington (1902–62) and Scott Russell Sanders (b. 1945). Chapter 3, "To Live in the Wilderness as a Wild Creature Myself," explores Errington's contribution as one of several midwestern wildlife biologists, including Aldo Leopold, who were in the vanguard of ecological thinking during the first half of the twentieth century. His theories about interdependence of wetland species were grounded in a deep, lifelong fascination with landscapes of the Upper Midwest, particularly Minnesota and the Dakotas; as a literary naturalist, he also indulged in lyrical introspection, what might be seen as his pastoral turn. Above all, he fashioned a vision of conservation rooted in an ecologist's deep regard for the workings of nature, and advocated preserving the integrity of natural ecosystems for posterity, as a source of solace and inspiration as well as for scientific inquiry.

Chapter 4, "The Wilds That Gave Us Birth," addresses the work of Scott Russell Sanders (b. 1945), who ranks among the foremost contemporary midwestern literary naturalists who carry on the legacy of the region's pioneering ecologists. He recounts how the region's astonishing abundance established over millennia had been largely exhausted in less than two centuries and posits the corresponding sense of ethical responsibility for protecting remaining pockets of ecological integrity. Moreover, he views environmental health and social justice as inseparable. In what ways, he asks, might we revise the land ethic today in light of the cultural legacy perpetuated by the continent's indigenous peoples? Sanders believes unequivocally that preservation of traditional cultures, languages, and lifeways is part and parcel of protecting the environmental quality.

The third pair considers Theodore Roethke (1908–63) with William Stafford (1914–93), twentieth-century poets who express ecological sensibilities as well as a fascination for the legacy of Native American cultures. Chapter 5, "Poetics of Place," explores how growing up in Michigan, Roethke gained an appreciation for the region's environmental history and an understanding of how development had transformed the landscape. He realized that within two generations the face of Michigan's woodlands had changed irrevocably, exacting a terrible ecological cost. He situated his interpretation of landscape change in the grand sweep of American history, with the Upper Midwest being settled by Europeans, and repeatedly invoked the figure of Indians in wilderness settings—an image conventionally associated with the "pristine" and primordial—as an homage to the continent's indigenous peoples. Accordingly, this chapter also examines how the legacy of Native American history and culture are represented in his work.

Chapter 6, "Landscapes of the Past," examines how poet William Stafford often incorporated elements of environmental history into his work. His attunement to nature and fascination with Indian cultures were originally forged growing up on the Great Plains in Kansas. The grasslands of his youth was a region where agricultural acreage intersected with vast tracts of open space, and it had a storied past: not the intact ecosystems of remote wilderness preserves, perhaps, but landscapes historicized by the arrival of white settlers. Stafford's many references to the indigenous peoples of North America—both their presence and absence—reflect an overall historical sensibility as well as a philosophical stance, embracing an ethical relationship to the natural world. He ascribed particular importance to the symbolic significance of Indians in the popular imagination: associated with a pristine wilderness that they were imagined to have once inhabited, Native Americans have commonly been romanticized as stewards of the wild.

The fourth set of chapters addresses contemporary Native American literature, pairing Ojibwe novelist Louise Erdrich (b. 1954) and Cherokee author Diane Glancy (b. 1941). Instead of giving a general survey of their work, I examine texts that depict how Native American languages are connected to the places where they historically evolved. Chapter 7, "Landscape and Language," considers Erdrich, above all her exploration of how indigenous languages and spiritual practices reveal an age-old relationship to nature. Rather than revisiting her best-known titles, this chapter focuses on a volume of nonfiction, *Books and Islands in Ojibwe Country* (2003). This literary travelogue recounts a personal pilgrimage to traditional tribal homelands in southern Ontario, Canada. Describing her journey by canoe through the Boundary Waters at Lake of the Woods, she contemplates how Ojibwe traditions embody an enduring relationship to particular ecosystems. Moreover, she contends that the legacy of that interaction is directly reflected by traditional languages. Learning Ojibwemowin in situ, she discovers, involves recognizing the plant and animal species that long sustained the Ojibwe people, and that still provide a basis for their spiritual practices. Place, she concludes, forms the bedrock of cultural identity. These meditations are situated at the confluence of cultural and ecological memory, offering a unique glimpse of the connection between Native American peoples and the landscapes they traditionally inhabited.

Chapter 8, "Giving Voice to History," considers Native American poet and novelist Diane Glancy's exploration of the relationship between indigenous peoples and place in a pair of linked novels, *Pushing the Bear: A Novel of the Trail of Tears* (1996) and *Pushing the Bear: After the Trail of Tears* (2009). This chapter examines both books from an ecocritical perspective to illuminate indigenous relationships to nature as well as the essential role of place in the formation and perpetuation of cultural identity. Glancy's account of relocation to Indian Territory along the Trail of Tears resounds with cultural memory of tribal homelands lost to removal. Taken together, these two novels offer an intimate glimpse of what was at stake for the Cherokee when they were forced to abandon their ancestral homelands: sacrificing a historical legacy of cultural and spiritual significance. She traces connections between language and landscape much like those Louise Erdrich contemplates in *Books and Islands in Ojibwe Country*, recognizing that Native American languages are intrinsically connected to the places where they evolved. For the Cherokee, as for other indigenous peoples of North America, environmental history and cultural identity have been, and always will be, inseparable.

The fifth and final pair of chapters juxtaposes two midwestern literary naturalists, essayists Elizabeth Dodd (b. 1962) and Paul Gruchow (1947–2004). Chapter 9, "Time's Horizon," traces how Dodd has emerged over the last two decades as a major voice, publishing three volumes of essays as well as two collections of poems that form a cohesive body of work probing relationship to place. In the tradition of Loren Eiseley, her writings exemplify an interdisciplinary stance as a naturalist, deftly shifting focus between anthropology, ecology, geology, and linguistics. Attuned to environmental history when afield, she contemplates past accounts of landscape change, ultimately attempting to square historical records of midwestern forests preceding white settlement with the second- and third-growth woodlands common in the Appalachian foothills she wandered as a child. Similarly, she reflects on the isolated remnants of primordial forest still standing in our time that provide a glimpse of ecosystems that once thrived throughout the region.

Chapter 10, "Islands of Time," examines Paul Gruchow's contribution as a literary naturalist writing at the cusp of ecology and memory, nature and culture. While the influence of Henry David Thoreau is pervasive in his thinking, his reverence toward the natural world and sheer lyricism might also be likened to John Muir's. Inspired in part by twentieth-century literary naturalists such as Loren Eiseley, Rachel Carson, and E. O. Wilson, his work follows in the broad tradition of American natural history writing by exploring nature in relation to multiple dimensions of time—whether diurnal, seasonal, geological, or evolutionary. In the wake of pervasive agricultural practices that had radically altered the grasslands, he recognized how the region had in fact become a crucible at the middle of the twentieth century for pioneering ecological thought and advocated preservation of representative examples of a variety of naturally occurring ecosystems for posterity.

The volume culminates with a brief epilogue making a personal plea for cultivating a culture of restraint, as well as considering the sorts of conservation initiatives supported by the writings of the authors discussed in this study, particularly in light of their reading of environmental history in the Midwest.

NOTES

1. Sanders, *Staying Put*, 114.
2. Gruchow, *The Uses of Wildness*, 10.
3. While the term *ecological memory* has been employed in a wide variety of disciplinary contexts ranging from landscape architecture and restoration to ethnobotany (see Nazarea, "Local Knowledge"), it is intended here in a more generalized descriptive sense corresponding to narratives of environmental history.
4. Nazarea, "Local Knowledge and Memory," 317–35.
5. Matthiesen, *Wildlife in America*, cover.
6. Lopez, *Of Wolves and Men*, 14.
7. Ibid., 70.

8. Ibid., 15.
9. Berry, *Unsettling of America.*
10. Ibid., 29.
11. Ibid., 30.
12. Van Noy, *Surveying the Interior*, xviii.
13. Ibid., 3.
14. Ibid., 7.
15. Ibid., 37.
16. Francaviglia, *Believing in Place*, xvii.
17. For a full discussion of these terms, see Lynch, "Ecopastoralism."

Reimagining Environmental History

Prologue

Opening with Thoreau

Rediscovering the wild
may ultimately be an act of discovery,
but initially and fundamentally
it will be a process of collective memory.

—Terry Osborne

On a midwinter day in January 1855, Henry David Thoreau noted in his journal that he had been reading environmental history again. On this occasion, he was perusing an early seventeenth-century account of New England's natural resources published by William Wood. *New England's Prospect* was the sort of tract intended to beguile potential colonists with images of fabulous abundance—the fabled bounty of the New World—while tantalizing investors who might underwrite the requisite ships for shuttling settlers and then carrying back furs, commercial cargo, including raw materials such as timber, and delicacies such as barrels of pickled fish, whose subsistence value for settlers had to be weighed against their commercial worth as exports. Literature promoting early British settlement was prone to making extravagant and dubious claims; by contrast, Wood's tone was considerably more measured—at moments even literary—than the exaggerated accounts of exuberant, early boosters who often relied primarily on hearsay, having no firsthand experience traveling in the New World themselves. In fact, he would be credited with publishing "the earliest comprehensive record of New England's natural resources at the beginning of the European settlement."[1] Clearly, he viewed the project as revisionist precisely because he deemed so many previously published accounts to be "imperfect," "false," and even "scandalous."[2]

When Wood returned to England in 1633, European colonization in New England was still primarily along the coastline, and still so scant inland that not so much as a single village had yet been successfully established there. Consequently, he emphasized resources along the shore, such as fish that were unimaginably plentiful in proximity to existing settlements, including "an abundance of salmon, shad, and bass," as well as "multitudes" of alewives, banks of foot-long oysters choking the shoals, and sturgeon reaching up to eighteen feet in length that were routinely preserved by pickling for shipment to England.[3] Alewives, spawning in rivers in astonishing numbers, could profitably be processed as herring; he reported watching "ten thousand taken in two hours by two men without any weir."[4] The colonists feasted seasonally on any number of marine, anadromous, and freshwater species, ranging from mackerel to shad and eels. Striped bass, a particularly prized delicacy, arrived on annual runs along the coast in such abundance that a lone Indian or English angler could readily catch as many as fifty or sixty during a single high tide by fishing with a simple hook and line. The indigenous peoples of the region had long relied on an eclectic diet—hunting and fishing, gathering and planting—one that directly reflected and capitalized on the region's striking biodiversity, also recorded in *Prospect:* "In wintertime they have all manner of fowls of the water and of the land, and beasts of the land and water, pond-fish, with [herbs] and other roots, Indian beans, and clams. In the summer, they have all manner of seafish, with all sorts of berries."[5] Lobsters were so commonplace that Indians usually used them for bait when fishing, while many settlers actually thought them not worth eating at all. Indeed, "there is no country known," Wood boasted, "that yields more variety of fish winter and summer."[6] Given the remarkable diversity of species, these year-round fisheries held enormous potential for commercial export to European markets, especially considering the size of the fish then being caught. What struck Thoreau, reading centuries later, was the extent of wildlife decline since that time.

Thoreau began to contemplate the epic drop in game populations during the two intervening centuries, as William Cronon recounts in his classic "Changes in the Land," cataloging predators and prey extirpated from the region entirely. The list was alarmingly long, beginning with major mammals, including game that had undoubtedly served as a dietary staple for the first generations of colonists. These early settlers had already quickly engaged Native Americans in the fledgling fur trade. Wood reported that Indians had historically trapped for their own use bears, beavers, cats, deer, foxes, moose, otters, and wolves. By the 1630s, however, they were trading with the British, exchanging both furs and meat of beavers, otters, raccoons, wildcats, and wolves for European goods.

Without Wood's account, along with those of his seventeenth-century brethren, we would have no way of knowing the extent of wildlife population decline, let alone how many animals have vanished. Among the apex predators referred to in *Prospect* were cats, including a "grim-fac'd Ounce" known today as the lynx. While regional names for several species, including bobcat, cougar, mountain lion, panther, puma, and wildcat, had once been used somewhat interchangeably, the fact remains that by Thoreau's time many big cats had already been extirpated from much of their historic range. In addition to the cats, a host of other species had long since been hunted and trapped out of existence in the region. Black bears, for example, had been especially plentiful in Wood's time—more a nuisance when berry picking than an actual hazard to cattle or colonists. Bear meat had been regarded by settlers as surpassing even the most savory venison. Both bear and deer were gone well before the mid-nineteenth century, as well as moose, along with the wolves that had preyed on them all. Many furbearers, such as marten and beaver, had also been trapped out completely. Bird populations had plummeted too, due in large part to overharvesting, as Thoreau knew, by commercial market hunters who had already thinned once burgeoning flocks of passenger pigeons and heath cock so severely that both species would subsequently become extinct.

In Wood's time, the monumental flocks of passenger pigeons he described with wide-eyed wonder were still at their height: "I have seen them fly as if the airy regiment had been pigeons, seeing neither beginning or ending, length or breadth of these millions and millions . . . so they continued for four or five hours together."[7] In his estimation, even intensive hunting appeared neither to disrupt nor even make a perceptible dent in such migrations: "The shouting of people, the rattling of guns, a pelting of small shot could not drive them out of their course . . . the Indians fetch whole loads of them."[8] Other fowl were abundant as well. Turkeys in flocks of up to one hundred were especially plentiful. Meanwhile, market hunters steadily plied their trade, typically taking a dozen turkeys in a morning; the larger cocks commanded a premium price, twice that of hens. A wide variety of other species were also commonly available for sale in colonial markets, including cranes, ducks, heath cocks, partridges, swans, and teals, as well as at least three types of geese. Wood described the slaughter of waterfowl practiced by early colonists: both subsistence and market hunters gunned with abandon, killing in numbers that are staggering by today's standards. While individual hunters routinely returned from a single outing with fifty or sixty birds, Wood remained convinced that populations in the wild remained undiminished. To the contrary, it seemed as if they might even be increasing; indeed, he contended, "I have seen more living and dead [birds] the last year" than ever before.[9]

While *Prospect* acknowledged the intensity of subsistence activities by early colonists, Wood contended that wildlife showed little discernible difference in numbers as a result, and he found no evidence that any major species had begun to succumb under such hunting pressure. Nonetheless, his hostility toward predators, "ravenous" and "howling," reflected centuries if not millennia of European antipathy toward the wolf, which he accused of greatly reducing bear, deer, and moose—game that he considered indispensable to colonists.[10] "There would be more of them if it were not for the wolves. It would be a good change if the country had for every wolf a bear, upon the condition all the wolves were banished."[11] He assumed that hunting would only improve once wolves had been reduced or even eradicated entirely, as would soon be the case. In his estimation, the populations of game remained stable enough to be ample for subsequent waves of settlers, who could hope to rely indefinitely on such species for sustenance. He subscribed unabashedly to the widely held assumption of his time that the New World's abundance was virtually inexhaustible. Ironically, however, the eventual extirpation of the most highly prized game species would not be caused by natural predators such as the wolf, but by the settlers themselves.

Thoreau realized that since Wood's time even the aesthetic pleasure of seasonal change had been undercut considerably as a succession of species disappeared: "changes of fur and plumage [had] ushered in the spring and marked the other seasons of the year."[12] Major migrations had changed as populations of waterfowl waned. The very progression of seasons was no longer manifested in the same ways. In fact, many of the seasonal markers that Native Americans had relied on to guide the timing of their annual cycles of subsistence, he recognized, were gone as well. His laments seemed to deepen as he considered all that was at stake philosophically and spiritually—as well as environmentally—and took on an aesthetic cast, occasionally framed in intimately personal terms: "I take infinite pains to know all the phenomena of the spring, for instance thinking that I have here the entire poem, and then, to my chagrin, I hear that my ancestors have torn out many of the first leaves and grandest passage, and mutilated it in many places."[13]

Thoreau returned repeatedly in his journals to the theme of reading environmental history, examining the written record and comparing past reports with his own observations of wildlife populations still to be found in the environs of Concord—noting above all those animals that were by then absent. Beyond providing an inventory of historical biodiversity in the region, he recounted from memory the litany of species lost: "The nobler animals have been exterminated here—the cougar, panther, lynx, wolverine, wolf, bear, moose, deer, the beaver, the turkey."[14] He understood that by his time the cumulative environmental decline had been devastating: "If salmon, shad, and alewives were pressing up our river now, as formerly they were… the very fishes in countless schools are driven out of a river by the *improvements* of the civilized man, as the pigeon and other fowl out of the air. I can hardly imagine a greater change produced by the influence of man on nature."[15] Indeed, he declared, it seemed almost as if "all the great trees and beasts, fishes and fowl are gone."[16] The countryside stripped of indigenous wildlife to such an extent he deemed to be "tamed," "imperfect," and even "maimed."[17] In the process, even the *memory* of previous abundance had been effectively obliterated. He concluded that we have inherited at best an "imperfect… notion" of "the actual condition of the place where we dwell, three centuries ago."[18]

By interspersing accounts written in previous centuries with his own observations in the 1850s, Thoreau established not only a metric for ecological memory, but an approach to reading environmental history by triangulating between past depictions of abundance and present observations, a practice perpetuated by literary naturalists to the present day, including several discussed in coming chapters, such as John James Audubon, Scott Russell Sanders, Elizabeth Dodd, and Paul Gruchow.

NOTES

1. Osborne, review of *Rediscovery of the Wild.* Vaughn, introduction to *New England's Prospect*, 8.
2. Ibid., 20.
3. Homan, *Yearning for Wildness*, 90.
4. Wood, *New England's Prospect*, 63.
5. Ibid., 86.
6. Ibid., 53.
7. Ibid., 50.

8. Ibid.
9. Ibid., 53.
10. Ibid., 42.
11. Ibid., 43.
12. Thoreau, *Journal of Henry D. Thoreau*, March 23, 1856. Most of the Houghton Mifflin 1892 volumes of the *Journal* are accessible for free through Google Books (books.google.com).
13. Ibid.
14. Ibid.
15. Ibid., April 11, 1856.
16. Ibid., March 23, 1856.
17. Ibid.
18. Ibid., January 29, 1856.

1

The Making of a Conservationist

JOHN JAMES AUDUBON (1785–1851)

You can tell where a mockingbird lives
by the songs he sings.

—JENNIFER ACKERMAN, *The Genius of Birds*

Best remembered for his precise and highly expressive paintings of wildlife, perhaps John James Audubon's most profound legacy, beyond a prodigious body of artwork and publications, is the ideal of conservation itself. While his achievements as consummate artist, accomplished naturalist, and aspiring entrepreneur are widely recognized, his contributions as author and conservationist remain less fully appreciated. His writings are certainly less well known, yet as author and environmentalist Scott Russell Sanders points out, "During the years when he was feverishly painting the illustrations that made him famous, he was equally busy writing journals, letters, autobiographical essays, and volumes of natural history. Taken together, these pages summon up for us the retreating wilderness."[1] His later writings depicting the impact of frontier settlement across North America in the first half of the nineteenth century clearly reveal a "growing ecological anxiety," though he had actually long recognized and lamented the wholesale destruction of wildlife from overhunting and destruction of habitat.[2] While later writings published posthumously, such as *The Quadrupeds of North America* and the *Missouri River Journals*, show a growing concern about declining wildlife, Audubon had already voiced such themes repeatedly in earlier journals kept throughout the several decades he was engaged as a field naturalist exploring unsettled territory in proximity to America's westering frontier in service of his painting and publications.[3] While he did not live to see the publication of George Perkins Marsh's seminal treatise on environmental history, *Man and Nature*, in 1864, he had independently reached similar conclusions about threats to natural abundance in the New World several decades earlier.

Two commonly held and seemingly incompatible interpretations of Audubon's legacy arise from reading his exploits in terms of *today's* ecological imperatives. The first view holds that as a profligate hunter himself, he forfeits any credibility with regard to environmental ethics. Up until his time, however, diminishing wildlife populations were viewed as a merely local phenomenon, and what by today's standards would constitute excessive hunting was then commonplace. Yet in the course of conducting those collecting expeditions, he chronicled major changes to game populations as ranges waned throughout North America. Moreover, he was among the first to recognize and articulate the imminent prospect of species extinction; accordingly, the countervailing position hails him as a prophet of conservation. In this view, his work represents the seminal contribution of a quintessential nineteenth-century naturalist, and is perpetuated to the present day by the environmental advocacy of his namesake Audubon Society, whether through habitat preservation for the protection of endangered species, or by championing species biodiversity. Underlying both is our difficulty in imagining wildlife populations that so dramatically exceed today's, frequently by an order of magnitude or more. His writings from the first half of the nineteenth century continue to provide a baseline for the abundance then still found across much of the continent. As a naturalist exploring the expansiveness of the American frontier firsthand, Audubon recognized the beginnings of its demise.

Surely it would be hyperbole to beatify Audubon as the "patron saint" of modern conservation in America. Yet not only did he display remarkable artistry when depicting birds and muster an astonishing drive to canvas the continent on a succession of collecting expeditions, but he exhibited a prescient sense of environmental history. In light of the most recent edition of the *Missouri River Journals* published in 2016, however, we must reconcile the picture of Audubon as a conservationist coming to grips with frontier development with his lifelong reputation as an avid sport hunter. With few exceptions, the tension between his views as a conservationist and passion for gunning, above all for birds—for food, for sport, and for professional use as an artist and naturalist—has generally been rationalized, downplayed, glossed over, or excised entirely by biographers.

Moreover, the credibility of some evidence routinely cited to demonstrate Audubon's views on conservation has been challenged. There is evidence that the text of several manuscripts may have been reworked after his death by editors, including both co-author John Bachman for *The Quadrupeds of North America*, first published 1851–54, and particularly the version of the *Missouri River Journals* that his granddaughter Maria Rebecca Audubon edited in 1897. In fact, several passages purportedly drawn from those journals expressing conservationist sensibilities explicitly and forcefully were published posthumously, and therefore were susceptible to tampering, potentially compromising their authenticity. The degree to which specific language was actually rewritten or even manufactured is sometimes difficult to establish definitively, given the gaps in the original handwritten manuscripts—comprising just over a third of the journey—that survive today.

Daniel Patterson's exhaustive examination, critique, and commentary on the *Missouri River Journals* traces the invention and reception of Audubon's reputation as a conservationist by examining textual integrity and how his views on conservation have been represented in book-length biographies spanning over a century. Biographers writing from 1901 to 2004 have vacillated between accepting and questioning the legitimacy of certain published proclamations—particularly those lamenting the potential extinction of bison—since in some cases it remains unclear whether the passages were written by Audubon. In Patterson's estimation, biographers have been inclined to interpret Audubon's views and conservationist legacy in light of environmental thinking in their own times. In the 1890s, for instance, when the earliest edition of the *Journals* was prepared, widespread public awareness of wildlife decline—above all among the vast herds of buffalo and astonishing flocks of passenger pigeons—had led to growing concern about possible extinctions. This raises the suspicion that such sentiments might have been incorporated into Audubon's writing by others at a later date. Similarly, biographers working in the 1960s were writing during the emerging environmental movement inspired in part by Rachel Carson's *Silent Spring* and reflected by political events such as the first Earth Day and passage of the Clean Air Act. Three biographies were published in 2004, the bicentennial of his arrival in the United States. Each echoes and perpetuates the long-standing, conventional narrative of Audubon's gradual awakening to conservation, heightened by revelations late in his life, especially on the Missouri River expedition, sometimes portraying the awakening dramatically as a kind of conversion experience. Even if there was a degree of mythmaking involved in depicting him as a legendary figure, Patterson himself ultimately concedes that "Audubon is one of the earliest voices for restraint on the land . . . as well as the earliest advocate for the ethical considerability of birds in this country."[4]

Audubon recognized the striking contrast between the remaining verdant wilderness of America and exhausted regions of the Old World. He was in fact well acquainted with the "civilized" landscapes of Europe. As a child, he had been brought up in France, and as an adult he spent more than a decade in the British Isles while arranging publication of *Birds of America.* Promoting that monumental project abroad, he intuitively maneuvered through the cultural riptides of a colonial era to achieve his own audacious aims. Catering to the expectations of European aristocracy about the New World, he portrayed himself as America's native son, despite his birth in Haiti and upbringing in France. The ploy of presenting himself as a frontiersman was ultimately successful in bolstering his image as a naturalist intimately familiar with the wildlife of America. To have taken on trappings of the frontier—and to have paraded that way before Europe's high society—was to invoke a myth in which the New World was seen as a kind of wild Eden. He had become its self-styled, and self-appointed, emissary. We can only imagine how flamboyant a "long-haired, wolfskin-clad American woodsman" would have seemed in nineteenth-century London.[5] Audubon positioned himself as an archetypal American pioneer through rustic attire and outlandish backwoods antics, "playing the woodsman's role in the drawing rooms of England with gusto and self-irony, imitating bird calls and Indian yells."[6] Since his arrival in England serendipitously coincided with publication of James Fenimore Cooper's *The Last of the Mohicans,* Audubon appeared to be the very embodiment of the New World romanticized in such fiction and embraced by the European imagination. As Audubon scholar Alton Lindsey puts it:

> During the years of promoting subscriptions to *Birds of America* among wealthy potential patrons, Audubon saw the necessity for self promotion. . . . He not only created visual images of wild nature, but served also as writer, director, producer, narrator, advertising manager and salesman. . . . It may have galled such a man to play the role of long-haired, wolfskin-clad American woodsman in the salons of Europe. Or perhaps, with his penchant for practical jokes, he enjoyed the irony. . . . [Subsequently] Audubon has come to symbolize both woodsman *and* conservationist, two somewhat

> *antithetical* types, in his adopted country. Both kinds of symbolism have always had a powerful grip on the American imagination, and today we realize that further loss of wilderness not only means the loss of a part of our past, but of a part of our future as well.[7]

Biographers writing more than a century and a half after he lived continued to depict him in just the way he would have wished, clad in the trappings of the "American Woodsman" in a transparent attempt to establish authority as a well-traveled field naturalist on the mercurial American frontier. As biographer Shirley Streshinsky suggests, with Romantic overtones, Audubon "had traveled into the deep American forests wearing deerskin leather stockings and moccasins; he had hunted with the Indian, had witnessed the beginning of the destruction of the wilderness."[8] Perhaps his many self-conscious attempts to embody the quintessentially American frontiersman were intended not only for the benefit of his contemporaries but for posterity as well.

On completion of *Birds of America* in 1839, Audubon proclaimed in the introduction to the fifth and final volume of *Ornithological Biography*, "I find my journeys all finished, my anxieties vanished, my mission accomplished."[9] That summer he settled his affairs in Europe and sailed back across the Atlantic for the twelfth, and last, time to America, where one great adventure still awaited him. He soon proposed yet another monumental project: *Viviparous Quadrupeds of North America*. As driven as ever, Audubon began to contemplate a long-deferred dream: an expedition to the wilds of the American West to sketch and collect specimens of as many mammals as possible. According to Streshinsky, "He yearned to venture into that West, still largely unexplored land beyond the Mississippi, the Indian lands, where few white men went. He dreamed of the Rockies, of Oregon, of California. There had never been the time, never the money, never a safe place to leave his family. Now he had all three, as well as a reason to go: research for the *Quadrupeds*. He made up his mind. He would go on 'this grand and Last Journey I intend to make as a Naturalist.'"[10] Despite having postured for decades as a frontiersman, Audubon had not actually ventured west of the Mississippi until this last collecting expedition, heading up the Missouri River into Indian Country in 1843. The timing of the expedition coincided with a remarkable moment in frontier history. Given that "an apparently endless supply of open land in the West had focused American attention on the wilderness, Audubon, as the 'American Woodsman,' seemed to personify all these interests in his bird and mammal paintings."[11]

Audubon's expedition on the Missouri at last allowed him to pursue a lifelong ambition to explore the western frontier. When returning from London to the New World twelve years earlier in 1831, he had already formulated a goal for subsequent exploration that in the end proved to be unattainable: "He aimed first to visit the Florida peninsula and the islands off its east coast, then to explore the country west of the Mississippi, before going on to Mexico, and if possible, continuing as far as California, crossing the Rocky Mountains and following the Columbia River to its mouth."[12] While such ambitious plans—reminiscent of the Corps of Discovery led by Lewis and Clark—ultimately exceeded his reach, in the course of his lifetime he completed an extraordinary range of collecting expeditions that spanned much of North America, floating the Mississippi twice in the 1820s alone, and later reaching the maritime provinces of present-day Canada. Yet even such extensive adventures could not have prepared him for the tumultuous change he would encounter on the upper Missouri.

How exactly did such experiences as an artist-naturalist in the field lead Audubon to arrive at his convictions regarding the preservation of wildlife populations and the natural landscapes that sustained them? Studying birds and mammals throughout much of North America, he recognized how initial European settlement and subsequent development were adversely affecting habitat. He realized how unmitigated destruction of forests and watersheds threatened game to the point of extirpation locally or even absolute extinction of species. In fact, understanding the relationship between habitat loss and wildlife decline kindled his sensibilities as a naturalist and inspired a nascent interest in conservation. What began as a quest to celebrate the natural abundance of North America by comprehensively documenting every species of bird took on new dimensions; recognizing the extent of development and the resulting depletion of game, he expressed deepening concern about the survival of species of birds whose range was rapidly waning, as well as about major game—above all the buffalo—succumbing to relentless overhunting by market gunners. Observing the epic decline of wildlife then under way awakened his dismay about rampant landscape change and ultimately shaped "his perceptions of a wild America in the process of being domesticated."[13] Audubon's extensive travels collecting specimens and documenting wildlife gave him a unique perspective on how frontier development quickly extirpated game and destroyed habitat. In search of specimens throughout his career, he had effectively "followed the military frontier into the species-rich environments," initially in the southeast, but then progressively westward.[14] He lamented widespread decline in the range of many species and destruction of their historic breeding grounds, as well as declining populations caused by a combination of overhunting and habitat loss. He was hardly alone in perceiving the depletion of game, since declines in wildlife were immediately evident to trappers and traders as well as indigenous peoples dependent on the same species for subsistence. Famed frontier Indian artist George Catlin, in fact, had called for setting aside a large territory for both wildlife and the "wild tribes" of Indians still thriving on the western frontier in the 1830s. To many, the tumult of development on the frontier had already come to seem inevitable.

Audubon appreciated what stood to be lost: "Through his eyes we see the swirl and ferment of frontier America; we meet a huge cast of characters, bird and beast, and human, glimpsed during that once-on-a-continent moment when all wilderness east of the Rocky Mountains was on the point of vanishing."[15] He recognized the extent of declines in populations of many species during his lifetime, and chronicled these in his writing through capsule narratives of environmental history that reflected a deepening appreciation for the scope and possible consequences of landscape change associated with settlement. He repeatedly depicted with trepidation the destruction of forests and fisheries, as well as the shrinking ranges of species, including several now extinct, such as the passenger pigeon, Carolina parakeet, and ivory-billed woodpecker. While Audubon's later works expressed increasing urgency about ensuring the survival of wildlife, this chapter explores how he had in fact voiced similar concerns many times before. After considering the account of his last collecting expedition on the upper Missouri—often cited as evidence that he increasingly embraced what today might be termed environmental sensibilities—I will examine earlier writings to determine how this perspective had actually developed throughout his career.

During his final collecting expedition on the upper Missouri in 1843, Audubon witnessed tumultuous destruction to both indigenous peoples and wildlife on the western frontier. The *Missouri River Journals* reflect that reverberations of westward settlement were by that time already being felt far upstream. His accounts of that journey, subsequently incorporated into *Quadrupeds of North America*, foretold and even "mourned the destruction of wild America."[16] On this trip he witnessed a number of profound changes in progress—the arrival of white settlers, the removal of Indians, and the epic decline of wildlife—that in aggregate left him unable to reconcile popular conceptions of the frontier West with what he actually saw.

Audubon visited the region on the heels of a number of illustrious visitors, such as explorers Meriwether Lewis and William Clark and European adventurer Prince Maximilian of Wied-Neuwied, accompanied by the artist Karl Bodmer. As art historian Ron Tyler notes, "Beginning in 1839, a spate of beautiful and exotic materials on the West began to appear. Karl Bodmer's stunning aquatint engravings accompanying Prince Maximilian's 1839 text of his journey up the Missouri River surely whetted Audubon's appetite even further, as did George Catlin."[17] By the time he arrived in 1843, the glorious portraits of Native Americans by previous visitors—even the most recent ones—bore little resemblance to the tribes decimated by smallpox that Audubon would describe in his own journal. His trip in the wake of several major epidemics made him an eyewitness to a catastrophe. There was of course no way to anticipate such a sudden change, particularly in light of the glowing accounts Catlin had published just a few years before. He was certainly aware of those who had preceded him: "Even though Audubon desperately wished that the area had not 'been trodden by White man previously' and tried to convince others of its virginal state, his frequent reference to George Catlin's books and his familiarity with other travel literature of the period makes it clear that he never really believed this to be the case."[18] In fact, "one of the main goals of the Romantic explorers was to find a pristine wilderness and Indian culture untainted by civilization. In fact, if that was one of Audubon's goals, he was too late, for Euro-Americans had been traveling among and trading with the Indians of the upper Missouri for more than a century."[19] "When George Catlin went west in 1830, the average easterner and the interested European had only a vague and confused impression of the country beyond the Mississippi and the people who lived there."[20] In a sense, Catlin had a blank slate: It had been only a few years, after all, since the United States had adopted a policy to move eastern Indians across the Mississippi in 1825. His accounts as a correspondent on the frontier were published, appearing in New York newspapers such as the *Spectator* and the *Daily Commercial Advertiser*, and were later reshaped into the two-volume *Letters and Notes on the Manners, Customs, and Conditions of North American Indians*, self-published in London in 1841.

Since Audubon's journey on the upper Missouri followed in the wake of Catlin's celebrated travels there, he understandably depicted Native Americans on the frontier in the *Missouri River Journals* and *The Quadrupeds of North America* with reference to his predecessor's accounts. Catlin's legacy includes many paintings commonly viewed as "historical documents" from a time when "Indians were [still] as independent as their aboriginal ancestors had been when they met the first white explorers. The Great Plains were still Indian country."[21] Catlin had traveled up the Missouri River early—as far as Fort Union at the mouth of the Yellowstone in what is now North Dakota in 1832. So artists and ethnographers visiting there afterward would inevitably return to his images of the Blackfoot, Assiniboine, Ojibwa, Crow, and other tribes for a glimpse of how things once were. Yet, as Ewers has observed, Catlin became "a controversial figure in American art for generations," accused by some of romanticism or even sensationalism. In fact, what accounts for his subjects' apparent glamour, opulence, and stately bearing is that he chose to depict the most elite and wealthy among each tribe.[22] Undoubtedly Catlin's celebrated "Indian Gallery" displayed in New York City in 1837 had heightened Audubon's anticipation of witnessing the grandeur of Native American peoples by traveling to the Western frontier himself.

Carrying Catlin's books with him on the expedition in 1843, Audubon naturally compared his own impressions to those of his rival who had visited the region only eleven years earlier, revealing the intensity, extent, and pace of change under way on the upper Missouri. In this respect, Audubon's approach to reading environmental history parallels that of Thoreau, who had consulted accounts of New England's natural resources published in the early seventeenth century to measure subsequent change. Audubon was actually writing just over a decade before Thoreau's journal entries of 1855, yet he already decried ongoing extirpation of wildlife on the still relatively unsettled western frontier and recognized the prospect of absolute extinction of key species. Moreover, while Thoreau's account captured wildlife decline and regional extirpation spanning *centuries* since initial settlement by colonists, Audubon rendered the compression of such forces on the upper Missouri that he personally witnessed during the middle of the nineteenth century—a catastrophe exacerbated by excesses of the fur trade as well as the rampant displacement of Native Americans from ancestral homelands further east. Indeed, by the 1840s, visions of a pristine western wilderness had to be reconciled with a rapidly changing landscape in which wildlife and Native American peoples in the region were both in precipitous decline. Audubon found the western frontier profoundly altered from what Catlin had encountered just over a decade before: "Catlin saw the prairie and its Indians in his youth, and he saw them before they had much contact with civilization. Audubon saw people already dispirited, corrupted, diseased, drunken, beggared, prostituted. And he saw the prairie with the eyes of one fast aging."[23] Consequently, even the most recently published accounts proved to be so outdated that they now seemed greatly exaggerated. Indeed, by 1843, any hopes of encountering the vibrant, "wild" tribes Catlin described had been dashed.

Like others who pondered the transformation of the western frontier between 1830 and 1860, Audubon was initially at a loss to comprehend the extent and pace of such change. As environmental journalist Michael Harwood observes in *On the Road with John James Audubon*:

> To be fair, no man knew or could visualize exactly what was happening in fur country; events were moving too fast and haphazardly. The fur trade had revolutionized the life of the western Indians in the short space of four decades. So had the westward displacement of eastern and middle-western tribes. And in 1837—five years after Catlin visited—the smallpox had come north with the annual fur company boat and killed thousands of Indians, reducing some villages and tribes to only a few families.[24]

For Audubon, this expedition west clearly represented a personal pilgrimage. Encountering the staggering natural abundance of the Great Plains, he declared, "It is impossible to describe or even conceive the vast multitudes of these animals that exist even now and feed on these ocean-like prairies."[25] Despite marveling at the abundance of wildlife in the region, he already sensed the possibility of its eventual demise, much like the eradication of bison that he had witnessed farther east decades earlier. In addition, while viewing animal behavior that might animate his paintings, he sketched the natural habitat of each species. His knowledge of regional topography would ultimately be reflected in the finished prints of mammals in *Quadrupeds:* "just as with the *Birds*, the landscapes played a significant role in the compositions: they told something of the animal's habitat."[26] Audubon's concerns about declining wildlife were nowhere more forcefully expressed than in the *Missouri River Journals*. By 1843, as biographer Richard Rhodes explains, "the eastern wilderness was filling up with new Americans, moving out in ever greater numbers from the tidewater. If there was wilderness yet to see, it was west, up the Missouri River to the Yellowstone River and beyond."[27] When he reached Fort Union at the mouth of the Yellowstone two thousand miles upstream from St. Louis on June 13, 1843, it was undeniably still the heart of what was popularly known as "Indian country."

This last collecting expedition of Audubon's storied career was devoted to what he and co-author John Bachman hoped would become a definitive work on the continent's mammals, to be titled *Viviparous Quadrupeds of North America*. Since the two believed that America's mammals had not yet been comprehensively addressed by any single work, they no doubt recognized what a significant contribution to the scientific literature such an endeavor was destined to become. Accordingly, traveling up the Missouri, he hoped to collect specimens and sketch virtually every species of mammal large and small east of the Rocky Mountains. The journey was essential to this project since a number of key species were unique to the West, including antelope, mountain sheep, and even the American bison by that time. As he would ultimately argue in 1846 in the introduction to his folio of American mammals, "many of the 'Quadrupeds of North America' were long since described by European authors, from stuffed specimens.... Comparatively little, however, has of late been accomplished toward the proper elucidation of the animals which inhabit fields, forests, fertile prairies, and mountainous regions of our widely-extended and diversified country."[28] Audubon described the project's intended geographical scope as encompassing the entire continent of North America, including regions he characterized as "at the present period an uncultivated and almost unexplored wild, roamed over by ferocious beasts and warlike Indian tribes."[29] While emphasizing the extent and difficulty of the enterprise, he also implied the inevitability of exploration and settlement.

In truth, during of the first half of the nineteenth century, much of what was thought of as "frontier" was in flux, as wave after wave of pioneers carved homesteads from wilderness: felling trees, clearing fields, breaking prairie sod, and planting crops. Especially in earlier years, Audubon was not above expressing enthusiasm about such settlement and progress, echoing the nation-building rhetoric of his time.[30] In many ways he embodied the competing discourses and values of an era when the limits of the seemingly inexhaustible abundance of North America had only begun to be exposed. While he was not alone in recognizing such limits, he used his position as an esteemed international figure to convey this concern to the broadest possible audience. The narrative of courageous pioneers settling on the frontier had enormous currency in his day, after all, and Audubon sought to sell books. Yet he clearly understood the dynamic and volatile process involved, as well as its escalating and increasingly detrimental impact on wildlife. He watched white settlement edging ever westward throughout his lifetime; moreover, he recognized how rapidly development could proceed in the wake of initial settlers.

Audubon's journals of the journey began as he wended his way from home in New York City to Louisville, where as a newlywed thirty years earlier he had lived for two years at the Indian Queen Hotel. The first letters he sent home in 1843 exclaimed at the area's subsequent expansion: "Louisville like Cincinnati has grown beyond all calculation.... Louisville is now an overgrown City."[31] Traveling down the Ohio River from Louisville, he boarded the dilapidated steamship *Gallant* bound for St. Louis. The boat soon passed Henderson, where he had spent a decade during the town's founding days as a prosperous merchant and speculator but which ended in financial ruin involving an ill-advised mill and a stolen steamboat he had once owned. By the time he left Henderson in 1819, he possessed only art supplies, a portfolio of bird paintings, and grand ambitions. In that initial portfolio were a number of works—such as the great horned owl and the sharp-shinned hawk—that already signaled the style of later paintings that would make him famous. Soon after leaving Henderson, he set out on the Ohio and Mississippi Rivers in 1821—the year that Missouri was granted statehood. The expedition proved to be a turning point, yielding some of the lifelike, animated depictions that were to become the hallmark of his mature style.

En route to the upper Missouri in 1843, Audubon marveled at how dramatically Henderson had grown in the decades since the 1820s and how even smaller villages were virtually unrecognizable due to progress: "We passed several villages now called cities and so improved in size that I with difficulty recognized them."[32] Evidence of the burgeoning population and migration westward was everywhere. His letters depict a whirlwind of activity on the docks of St. Louis: "horses on board, waggons, carts, carriages, and furniture of all description belonging to new settlers going to the Missouri frontiers."[33] This frantic migration westward is a backdrop to his depiction of the expedition as a whole. Indeed, rapid and dramatic transformation is central to his accounts of the frontier. He personally felt the full force—the vertigo—of such rapid landscape change. In fact, within a generation of his collecting expedition on the Missouri River, the isolation of even such remote regions would give way to the railroads: by 1869, the Union Pacific would complete the first transcontinental route across North America.

The Missouri is among the most storied watercourses in American history. From its headwaters at Lolo Pass on the present-day Idaho-Montana border to its confluence with the Mississippi north of St. Louis, at 2,464 miles the Missouri is the nation's longest river.[34] Moreover, when Lewis and Clark led the Corps of Discovery in search of a transcontinental waterway at the dawn of the nineteenth century, they chose the Missouri. This event—in conjunction with the Louisiana Purchase and President Jefferson's vision for American commerce in the Rocky Mountain West—anointed the river as the corridor of choice for waves of trappers and traders, soldiers, and settlers who followed in quick succession. Traveling there in 1843, Audubon observed a region already in the process of a perhaps irrevocable change.

Audubon described departure from St. Louis for the frontier with bravado: "Hie with us there to the West! Let us quit the busy streets of St. Louis, once considered the outpost of civilization but now a flourishing city, in the midst of a fertile and rapidly growing country, with towns and villages scattered for hundreds of miles beyond it."[35] As scholar Robert Peck recounts, "When John James Audubon and the rest of his party of five left the St. Louis docks on the morning of April 25, 1843, for a six-month trip up the Missouri, the self-proclaimed 'American Woodsman' was fulfilling a dream that had long possessed him."[36] As ecocritic John Knott concludes, "Beyond the frontier regions of settlements and squatters cabins lay what he saw as the primal American wilderness."[37]

For the first several hundred miles of the trip, Audubon described pioneers carving out homesteads, a precarious foothold only to be dislodged by flooding. "The Number of habitations halfway in the water is quite Surprising," he wrote;[38] "in a great number of instances the squatters, farmers or planters, as they may be called, are found to abandon their dwellings and make towards higher ground."[39] Writing from Boonville, Missouri, two hundred miles from St. Louis, Audubon described "settlements all along both shores and the Lands [are] rich enough to produce anything."[40] He realized that even remaining open space was destined for development and portrayed the region as already ripe for pioneers: a "truly beautiful site for a town or city, as will no doubt [arise] some fifty years hence."[41] The Missouri

River seemed a porous border between worlds, where the distant drumbeat of progress could already be heard.

Audubon visited the western frontier in an era when Indians were increasingly being pressed westward from traditional territories onto reservations as wave after wave of white pioneers, prospectors, and settlers advanced in their wake. Nonetheless, he wrote, "after we have passed Independence we will enter the Indian countries."[42] Just two weeks upstream of St. Louis at Council Bluffs, he breathlessly announced: "We are now positively out of the United States boundaries Westward and now all the people we will see will be Indians, Indians, and nothing but Indians."[43] He was keenly interested in encountering firsthand the "wild" bands that Catlin had described. Sailing two thousand miles to Fort Union from St. Louis in a record forty-eight days and seven hours, he had many opportunities to observe and interact with members of a variety of tribes, including the "Sac, Iowa, Fox, Potawatomi, Omaha, Ponca, Arikara, Mandan, Assiniboin, and Gros Ventres."[44]

In fact, from the moment Audubon boarded the steamer *Omega* on April 25, 1843, he had already been among Indians—fellow passengers embarking from St. Louis for Indian territories upstream. He wrote: "We have 7 or 8 Indians, males & females, returning to their wigwams about 500 miles up the River," whom he described as "poor souls" and "tranquil onlookers."[45] Their destination was a broad prairie where, according to the resident Indian agent Major Richardson, some twelve hundred Iowa and Fox were congregated in proximity to fur traders and troops. Throughout the first weeks of the journey, Audubon became acquainted with the patchwork of ambiguous landholdings and the ever-shifting proximity of Indians and whites. As early as 1843 white squatters had begun to settle as far up the Missouri as Fort Pierre. Edward Harris, a naturalist who accompanied Audubon up the Missouri, notes in his journal passing an established "Farm" near the fort, which possessed livestock as well as crops of potatoes and "Mandan corn" abundant enough to trade with travelers.[46] On May 16, he recounted passing "ten or more Indians who had a large log cabin and a field under fence" immediately before the steamer's arrival at a rudimentary fort maintained by a fur-trading company at the Vermillion River.

When the *Omega* at last reached certifiable "Indian Country"—stretching to the west of the Missouri—Audubon observed a "good number of Indians" whom he characterized as "partly civilized."[47] Yet white settlers on their own scattershot homesteads were on hand to hail the steamer as it passed. As the steamboat approached the Black Snake Hills, he described the settlement and its proximity to Indian land: "There are a few houses, one or two Grist and Saw Mills, and an abundance of loafers *of all descriptions*. We have passed the Kickapoo's Nation and are now opposite to that of the Sacs." He added: "In 3 or 4 more days we will have passed the bounds of Civilisation."[48] He had ached for this moment, announcing in a letter written three days later that they had at last crossed "the line dividing the State of Missouri from the wilderness I am now pleased to call the Country ahead of us."[49]

Audubon realized, however, that even the most extensive tracts of unsettled territory were susceptible to development, adding a sense of gravity—perhaps even urgency—to chronicling wildlife in his time. An underlying premise of *Quadrupeds*, like the entire *Birds of America* enterprise, was to preserve for posterity a glimpse of the abundance and diversity of wildlife on the western frontier as it once had been. In the process, his own legacy became intertwined with the natural world that had shaped his sensibilities as a champion of wildlife.

Bachman urged him to keep copious and accurate notes of his observations of wildlife on the upper Missouri to establish professional credibility: "Everything...you write about them will be new. In a word give a true history of every species that inhabits the plains & mountains—the earth, rocks, & trees. This is the duty of the Naturalist."[50] While his account of the journey in the *Journals* dutifully cataloged specimens collected, birds sighted, and game taken as a matter of course, it also portrayed frontier communities and remnant Indian bands then undergoing dramatic transformation. Yet as Peck observes, it was in fact "Bachman's scientific rigor... [that] transformed a randomly organized aesthetic exercise and collection of anecdotal species accounts into a meaningful contribution to the collective knowledge of American natural history."[51] Even while striving for scientific authority in works such as *The Viviparous Quadrupeds of North America*, Audubon often indulged in seemingly serendipitous narratives as if to establish veracity based on his experience observing animals in their natural habitat.

Audubon understood that it took sustained observation to accurately characterize the behaviors of any species. His accounts of wildlife in the field—beyond reflecting an element of drama or romance—would lend the impression of direct personal experience. As Lawrence Buell points out: "*The Viviparous Quadrupeds of North America*, a book that Thoreau knew well, is a serious work of natural history intended to pass muster with zoologists. Certain portions of their entries are cut-and-dried: on description, color, dimensions, range. But the sections on 'habits' and 'general remarks' stretch out to what seems a delightful infinity."[52] At times such passages provide a vehicle for reflections on habitat loss and wildlife decline as well their impact on indigenous peoples. He clearly understood that incorporating such anecdotal material could make his books more appealing than the relatively lifeless recitation of scientific information alone. He painted birds actively engaged with their natural surroundings and later leavened his written descriptions with narrative accounts of his experiences afield, targeting a much wider audience in both the United States and abroad who shared a fascination with the wild landscapes, indigenous peoples, and ongoing exploration of the American West.

During the journey, Audubon marveled at the staggering natural abundance yet also recognized that wildlife was already being threatened with decimation, above all the buffalo. The range of the herds, native to the long-grass prairies of the Midwest during his own youth, had receded steadily westward. He lamented how the once abundant bison had disappeared from the Indiana and Illinois prairies. By the time he set out on foot in 1811 at twenty-five years of age, traversing the Illinois Territory for Henderson, Kentucky, he found the prairie already littered with skulls, horns, and bones of buffalo, a scene foreshadowing the carnage he would encounter out West on the upper Missouri three decades later.[53] He warned of their precipitous decline in his journal—a passage published posthumously in *Quadrupeds*, with slight rewriting by co-author John Bachman. Here's Audubon's original language (based on the Beinecke manuscript): "Their range has since that period gradually tended westward, and now you must take to Indian country, and travel many hundred miles beyond the fair vallies of the Ohio, towards the great Rocky chain which forms the back bone of North-America, before you can see the Buffalo feeding in his sturdy independence upon the vast elevated plains."[54] To witness such a spectacular scene, one must accompany him, at least figuratively, to the upper Missouri. By 1843, the great herds had already been driven that far west.

After recounting the many buffalo hunts he had witnessed on the journey, Audubon turned to how a variety of Indian tribes made use of the animal—as close as he ever got to ethnography. Detailing the strategies used by individual tribes, he contrasted the Mandan's pursuit on horseback, for example, with the buffalo jumps employed by the Gros Ventre, Blackfeet, and Assiniboine, who would stampede herds off a cliff. In the *Missouri River Journals*, he repeatedly chastised white hunters for taking only the tongue or hide and wasting the rest, whereas Indians reportedly found virtually every part of the buffalo to be valuable, whether for boots or bedding, shelter or sustenance. Scraps of hides were sewn into moccasins, hats, gloves, and even robes for children. Bones were ground to prepare marrow as a condiment. Bows were strung with buffalo gut. Shoulder blades served as primitive hoes. "Nothing is lost or wasted," he concluded, "but every part of the animal, by the skill and industry of the Indian, is rendered useful."[55] He decried the decimation of herds precisely because they were essential to the survival of indigenous peoples.

Based on his observations on the western frontier, Audubon contrasted the discretion of traditional subsistence practices with decimation of the herds by commercial market hunters. "Thousands multiplied by thousands of Buffaloes," he reported in his journals," "are murdered in senseless play, and their enormous carcasses are suffered to be the prey of the Wolf, the Raven, and the Buzzard."[56] Moreover, he scorned commercial hunters for decimating the buffalo herds serially. Indeed, he saw evidence of alarming exploitation, such as when a shipment from Fort Pierre passed the *Omega* on four barges with colorful frontier names such as "War Eagle," "White Cloud," "Crow-feather," and "Red-fish" loaded with ten thousand buffalo robes. He learned that these same boats would make several such hauls in a single season. He lamented in his journal: "What a terrible destruction of life, as it were for nothing, or next to it, as the tongues only were brought in, and the flesh of these fine animals was left to the beasts and birds of prey, or to rot on the spots where they fell. The prairies are literally *covered* with the skulls of the victims."[57] In fact, he was among the first to recognize and speak out about safeguarding wildlife: "If others thought the great herds would last forever, Audubon did not. In *Quadrupeds* he wrote that the buffalo was 'perhaps sooner to be forever lost than is generally supposed.'"[58] With a naturalist's eye, he viewed the decline of the bison as an emblem of the fate of the frontier as a whole.

Audubon simultaneously recognized the striking decline of "wild" Indian tribes. In fact, uncanny parallels had arisen between the plight of the continent's indigenous peoples and that of animal species. Like the buffalo, by the middle of the nineteenth century Indians were under siege. In certain ways their fates seemed tragically intertwined. Since the buffalo was central to their well-being, after all, its eradication would undermine a way of life. Accordingly, the ethical imperative of ensuring survival of game populations intersected directly with the protection of indigenous peoples that depended on these species. Above all, he recognized that Plains tribes relying so heavily on the buffalo—a *single* species—for food, clothing, and shelter were already being jeopardized. He realized that to eliminate the buffalo—a policy perpetrated by the US Cavalry under the likes of General William Tecumseh Sherman—was to imperil these peoples. If the herds were in effect the Indians' "commissary," eradicating the bison was tantamount to plundering planted fields or torching stores of corn. Faced with starvation, the tortured logic went, they would have little choice but to surrender.

Witnessing firsthand both the decimation of Indian nations by epidemics and dislocation in the wake of European exploration and settlement—exacerbated by the buffalo herds receding—forever altered his appraisal of the prospects and predicament of America's indigenous peoples. "Audubon believed that 'civilization' had been disastrous for the Indians; he was sickened by it, sorry for it, but like most men of his age, he accepted it as the inevitable toll taken by progress,"[59] a sentiment that might be characterized as a strain of fatalism: the assumption that for better or worse progress was destined to continue, accompanied by considerable collateral damage. The conventional wisdom of Audubon's day, after all, was that Native Americans were doomed. In fact, many of his contemporaries simply assumed that they were destined to vanish. To view them instead as the rightful inhabitants of ancestral homelands across the continent, as Catlin had, provided a counterpoint to the doctrine of Manifest Destiny that sought to legitimize westward expansion in the nineteenth century.

Audubon had expressed sympathy years before for the plight of the First Nation peoples of Labrador as well as the Creeks and Seminoles in the Southeast.[60] On a collecting expedition to the Gulf of Mexico in 1837, for instance, he had observed the forced relocation as several thousand were marched from their ancestral lands through Alabama under military escort, a precursor of the Cherokee Trail of Tears the following year. He was mortified by the sight.[61] At the very moment his adopted country was burgeoning westward, he recognized just how high the stakes could become for Native Americans and wildlife alike. He knew that in the span of a single lifetime the range of wildlife had been reduced dramatically, including timber wolves and grizzly bears, besides the iconic bison. The destruction of habitat by frontier development, coupled with the eradication of game, was an all too familiar chain of events: "Audubon was smelling the slaughter that three decades later would reduce the North American bison nearly to extinction. He may have connected the casual slaughter he saw with slaughter he had witnessed of passenger pigeons, which flocked in even greater numbers than the buffalo."[62]

Audubon returned from the Missouri somewhat disillusioned; clearly, he had arrived on the frontier just after the passing of what must have seemed to him a proverbial golden age. Neither the landscape nor its indigenous peoples had measured up to his lofty expectations. The "Wild West," it seemed, was not all it had been purported to be. He grew increasingly skeptical of earlier accounts—especially Catlin's—which he now believed had misled him by exaggeration if not outright fabrication. Moreover, he expressed bitterness for having arrived on the western frontier too late to observe the region in all its fabled glory.

Perhaps Audubon's most striking statements concerning the exploitation of game are to be found in later works such as the *Missouri River Journals*, and subsequently echoed in *Quadrupeds*. His dismay about overhunting and frontier development had clearly grown throughout his career, for he had expressed such sentiments repeatedly in previous writings. Yet how exactly had he come to recognize that habitat loss during his lifetime had led to decreases in the historical range of species?

As early as 1821 while floating the Mississippi River, Audubon described in his journal both the abundance of birds, "the passage of millions of golden plovers," and the alarming numbers gunned down in a single day: "the sportsmen are here more numerous and at the same time more expert at shooting on the wing... assembled in parties of from 20 to 1000 at different places where they knew by experience they told me the birds pass.... A man near where I was seated had killed 63 dozens—from the firing before & behind us I would suppose that 400 gunners were out. Supposing each man to have killed 30 dozens that day 144,000 must have been destroyed."[63] Assuming that such a scene was repeated at intervals along a migration flyway, it is easy to imagine how quickly entire populations of even the most numerous species might be threatened. As Harwood observes, while "the journal of 1820–21 contains the earliest surviving writing about birds," Audubon kept such journals of his expeditions and exploits for better than two decades, "filling them with observations and sketches, and he would eventually draw on this rich collection as he wrote the *Ornithological Biography*."[64] Decrying the "profligacy of frontier hunters" became a lifelong refrain.[65]

As biographer Duff Hart-Davis observes, "For a modern reader, some of the most haunting passages are those in which the author depicted the pristine nature of the wilderness along the Ohio when he first knew it, and lamented the destruction which was already taking place" by the 1820s.[66] "The Ohio," just such a nostalgic "episode" that appeared in the first volume of *Ornithological Biography* in 1831, reads like a Romantic lament for the trade-offs made in the name of *progress:*

> When I think of these times, and call back to mind the grandeur and beauty of those almost uninhabited shores…unmolested by the axe of the settler…when I see that no longer any Aborigines are to be found there, and that the vast herds of elks, deer and buffaloes which once pastured on these hills and in these valleys, making for themselves great roads to the several salt-springs, have ceased to exist…when I see the surplus population of Europe coming to assist the destruction of the forest, and transplanting civilization into its darkest recesses…a country that has been so rapidly forced to change her form and attire under the influence of increasing population.[67]

Even as a young man, Audubon's enthusiasm for protecting the natural world seemingly knew no bounds. As Robert Paxton recounts, "He knew enough of early industrialization in the United States to foresee the end of nature there. He imagined summoning Sir Walter Scott…to describe American nature" so as to celebrate wilderness still unspoiled and plead for its preservation.[68] In a journal he kept while living in England in 1826, Audubon had rehearsed what he hoped to communicate to Scott, whom he had not yet met, and whose support he evidently wished to enlist for commemorating nature in the New World and preserving America's wilderness. He undoubtedly craved the endorsement of one of the best-known and widely read novelists of his time. In that journal, Audubon declared: "Wrestle with mankind and stop their increasing ravages on Nature and describe her now for the sake of future ages. Neither this little stream, this swamp, this grand sheet of flowing water, nor these mountains will be seen a century hence as I see them now.… Fishes will no longer bask on the surface, the eagle scarce ever alight, and these millions of songsters will be drove away by man."[69] As Rhodes concludes, Audubon already "recognized that one consequence of this invasion was a decline in bird populations."[70] He was equally aware of threats to other wildlife as species after species succumbed to overhunting and destruction of habitat in various locales.

In the late 1820s, on a six-week collecting expedition in the Great Pine Swamp in northeastern Pennsylvania, Audubon watched with dismay the pace of logging operations felling the forest of the Lehigh River watershed. "Before population had greatly advanced in this part of Pennsylvania," he observed, "game of all descriptions found within that range was extremely abundant. The elk did not distain to browse on the shoulders of the mountains, near the Lehigh. Bears and the Common Deer must have been plentiful."[71] So too the turkeys, pheasant, and grouse. Yet already, rivers once teeming with fish were now clogged with rafts of logs floating to mills and downstream to lumber markets as distant as Philadelphia. Such intensive logging operations, he understood, would rapidly degrade the landscape, destroying the habitat for wildlife that provided sustenance for settlers dependent on such subsistence resources. He sensed a finality in such desecration, writing, "No sooner was the first sawmill erected than the axemen began their devastations. Trees one after another were, and are yet, constantly heard falling during the days; and in calm nights the greedy mills told the sad tale, that in a century the noble forest around should exist no more."[72] Similarly, when traveling through Maine en route to Labrador in 1833, Audubon again recoiled at the violence of such logging operations, describing how they could savage a watercourse: "Logs rolled, reared, tossed and tumbled amid the foam...emblem of wrack and ruin, destruction and chaotic strife...from out the frightful confusion of the scene came a melancholy feeling that did not entirely leave me for many days."[73] Moreover, he understood that the "virgin," old-growth forests being cleared would not regrow the same: "Whenever the first natural growth," he wrote in his journal during that expedition, "is destroyed by fire, or the axe, or by the hurricane, another of quite a different kind springs up."[74] He recognized not only that the younger trees of such second growth would be dwarfed in comparison to the original woodlands, but that conifers were typically replaced by a mixture of slower-growing hardwoods: a different kind of forest entirely.

Audubon's grasp of environmental history is evident from the first pages of the *Birds of America* portfolio, as well in the *Ornithological Biography*, which begins with an account of how the turkey's range was already beginning to dwindle, a cautionary tale of overhunting plus lost habitat. Consider the fact that he selected the turkey as the very first of 435 plates to be engraved and published. He sought to open each successive volume of *Birds* with a particularly striking image so as to heighten the anticipation of subscribers. The dramatic, life-sized likeness of the turkey would fit the bill.[75] Yet he had chosen this spectacular image to lead off the entire portfolio for another important reason: to explain changes that he had witnessed in its range in great geographic detail.

> The unsettled parts of the States of Ohio, Kentucky, Illinois, and Indiana, an immense extent of country to the north-west of these districts, upon the Mississippi and Missouri, and the vast regions drained by the rivers from their confluence to Louisiana, including the wooded parts of Arkansas, Tennessee, and Alabama, are the most abundantly supplied with this magnificent bird. It is less plentiful in Georgia and the Carolinas, becomes still scarcer in Virginia and Pennsylvania, and is now very rarely seen to the eastward of the last mentioned States. In the course of my rambles through Long Island, the State of New York, and the country around the lakes, I did not meet with a single individual.[76]

The turkey's diminishing range due to deforestation and overhunting provided a case study in the need for habitat preservation. "When the margins of our rivers shall have been drained and ploughed to the very tide mark… Turkeys shall no longer be met with in a wild state."[77] Such elegiac passages as this one from the first volume of *Ornithological Biography* offer a poignant lament for the destruction of habitat and its impact on wildlife. America's natural abundance, he realized, was already imperiled.

Audubon's extensive travels and lifelong devotion to studying birds allowed him to develop a perhaps unprecedented level of knowledge about many species. He was an extraordinarily attentive if largely self-taught naturalist, cited by his contemporaries, including Henry David Thoreau.[78] In fact, when Thoreau was beginning his publishing career in the early 1840s, twelve years before *Walden* appeared in 1854 and twenty-three before his final title, *Cape Cod* in 1865, he was already reading Audubon's writings with relish, declaring in *The Natural History of Massachusetts* (1842): "I read in Audubon with a thrill of delight, when the snow covers the ground, of the magnolia, and the Florida keys…of the breaking up of winter in Labrador, and the melting of the snow on the forks of the Missouri; and owe an accession of health to these reminiscences of luxuriant nature."[79] Similarly, Charles Darwin cited only one other naturalist more frequently than Audubon, mentioning him in several of his most influential works, including both *The Descent of Man* and *The Origin of Species*.[80] Because of his lack of formal education in the sciences, as Sarah Boehme observes, any claims of expertise were hard earned: "Audubon's projects depended on his experiences as a field naturalist.... He promoted the position that his authority rested on his extensive field experience."[81] His conviction that natural history is most trustworthy when derived from personal experience in the wild defied the typical practices of the elite scientists and "cabinet naturalists" of his time.

Given the virtually unparalleled extent of his fieldwork throughout North America, Audubon was uniquely positioned to describe the geographical extent of the breeding range of migratory species such as the Canada goose:

> It is found to breed sparingly at the present day, by many of the lakes, lagoons, and large streams or our Western Districts, on the Missouri, the Mississippi, the lower parts of the Ohio, on Lake Erie, the lakes farther north, and in several large pools situated in the interior of the eastern parts of the States of Massachusetts and Maine. As you advance farther toward the east and north, you find it breeding more abundantly. While on my way to Labrador, I found it in the Magdeleine Islands...and in Labrador these birds breed in every suitable marshy plain.[82]

He could also depict variations in habitat in considerable detail, having visited so many of these places himself: "for every hundred seen during the winter along our large bays and estuaries, as many thousands may be found in the interior of the country, where they frequent the large ponds, rivers, and wet savannahs."[83]

Similarly, Audubon could speak with remarkable particularity about variations in the timing and routes of major migrations:

> Its departure is always determined by the advance of the season, and the vast flocks that winter in the great savannahs or swampy prairies south-west of the Mississippi, such as exist in Opellousas, on the borders of the Arkansas River, or in the dismal 'EverGlades' of the Floridas, are often seen to take their flight, and steer their course northward, a month earlier...during the winter months, they are keenly sensible to changes of temperature, flying north or south in search of feeding-grounds.[84]

Moreover, he could delineate the *historic* range of a species and recognize how it had changed during his lifetime, incorporating earlier accounts such as General George Clark's stories reporting that seventy-five years earlier "wild geese were so plentiful at all seasons of the year, that he was in the habit of having them shot to feed his soldiers, then garrisoned near Vincennes, in the present State of Indiana.... The greater number [now] remove far north to breed."[85]

Several prints from *The Birds of America* folios depicting species lost to extinction have taken on profound environmental symbolism over time (and, ironically, remain among the images that collectors particularly prize), perhaps above all that of the passenger pigeon, whose flocks were unimaginably abundant in Audubon's day. Other naturalists and ornithologists of the time had marveled at their monumental migrations. Alexander Wilson, for instance, estimated a single flock flying over the Indiana Territory in 1810 at 2,230,000,000. Traveling between Henderson and Louisville in 1813, Audubon himself witnessed a flock that he numbered at over a billion birds. Positively awestruck, Audubon declared it "a torrent—and with a noise like thunder...that presently resembled the coils of a gigantic serpent."[86] Audubon's account goes on to suggest both the magnitude and the power of the spectacle: "But the birds poured onward in such multitudes that I rose, discouraged, after having recorded 163 flocks in twenty-one minutes. The farther I proceeded, the more I met. The air was filled with Pigeons that obscured the light of noonday like and [*sic*] eclipse."[87] Elsewhere, he likened the devastation of forests where such flocks roost to a natural disaster, "as if the forest had been swept by a tornado."[88]

Biographer William Souder recounts how Audubon marveled at flocks filling the sky from horizon to horizon, puzzling over how to calculate their numbers in the form of a daunting mathematical word problem. Audubon wrote,

> Imagine a column of passenger pigeons one mile wide, and assume this mile-wide flock passes overhead in three hours. If the birds are flying at 60 miles per hour, the whole flock could be visualized as occupying a rectangular area one mile wide and 180 miles long—180 square miles. Now assume a density of 2 birds per square yard...and you can figure the total number of birds at 1.1 billion.[89]

Given the dimensions of the flock described and the fact that Audubon was viewing just part of a still larger migration, their numbers may well have greatly exceeded his calculations. That such an abundant species could so swiftly become extinct must have seemed unthinkable.

Years later, Audubon recalled such a sight when traveling downstream from the Ohio's confluence with the Salt River: "Before sunset I reached Louisville.... The pigeons were still passing in undiminished numbers, and continued to do so for three days in succession.... The banks of the Ohio were crowded with men and boys, incessantly shooting at the pilgrims."[90] Such wanton slaughter, of course, eventually contributed to the population crash that ultimately led to their extinction. "On such occasions," he recounted in *Ornithological Biography*, "when the woods are filled with these Pigeons, they are killed in immense number, although no apparent diminution ensues."[91] He anticipated that the greatest threat was actually loss of habitat: "I have satisfied myself by long observation that nothing but the gradual diminution of our forests can accomplish their decrease."[92] As Audubon had lamented as early as 1831 in *Ornithological Biography*, the wild haunts he had wandered only decades before were already being lost to development. As Streshinsky concludes, "In less than a hundred years he would be proved right; as the woods were vastly diminished, the passenger pigeon became extinct."[93]

Similarly, the "Immense Flocks of Parakeets" that Audubon recorded in his early journals while floating the Mississippi in 1820–21 would be extinct within a hundred years.[94] As early as the 1830s, when his account of this species was published in *Ornithological Biography*, he recognized the population's decline: "along the Mississippi there is not now half the number that existed fifteen years ago."[95] While this passage did not explicitly attribute the cause to the impact of settlement, destruction of habitat, and extermination of flocks, Audubon did extrapolate from local extirpation to wider habitat decline: his "observation that they are very rapidly declining in number and contracting in range implies a sense of loss."[96] So too the ivory-billed woodpecker—another chapter in the saga of America's diminishing natural abundance. Both birds were still so plentiful in 1830 that Audubon had no qualms about collecting skins of the woodpeckers and parakeets while sailing from New Orleans to Louisville, as specimens for scientific and artistic purposes—demonstrating the seemingly paradoxical practice of nineteenth-century naturalists of "preserving" wildlife by shooting it first. This was standard operating procedure in an era predating photography: "To study the creatures of the wild he had to kill them... neither he nor anyone else thought of this as indiscriminate slaughter; that perception came much later, when the destruction of the forests and the wanton shooting of wildlife destroyed some species, and brought others close to extinction."[97] The assumption of inexhaustible abundance was yet to be widely questioned at "a time when vast flocks were commonplace and no one worried about limiting the number of birds shot."[98]

While retracing Audubon's expeditions in the 1970s, Michael Harwood realized that "without the assistance of a miracle we wouldn't find any of the strikingly handsome crow-sized ivory-billed woodpeckers. By contrast, in 1820 they were not only common in the Mississippi floodplain; in places they were the most evident, commonest woodpeckers."[99] Audubon had described the impressive extent of its historic range as once encompassing Kentucky, Indiana, the Carolinas, Georgia, Alabama, and Mississippi, as well as "all" densely forested tributaries west of the Mississippi River. The jury is still out on the status of this elusive species in the twenty-first century, as John Terborgh explains: "Even today, when extinctions are occurring with unprecedented frequency, they are extremely difficult to verify. The ivory-billed woodpecker is a prime example. The last substantiated photographs and sound recordings were made in the late 1940s, yet rumors about the bird's existence and fervent claims of sighting continue to emerge from the Southeast. Is the ivory-billed woodpecker extinct? No one can say."[100] Documenting a single extinction conclusively (let alone anticipating one) can be a vexing proposition indeed.

Audubon had even railed against the extermination of birds commonly considered to be pests by farmers and bounty hunters, such as the common crow. As late as the 1930s, government-sponsored eradication programs were still common, such as a single operation employing dynamite in Illinois credited with eliminating in excess of three hundred thousand crows.[101] Audubon's logic was unimpeachable: surely, he reasoned, "a portion may well be spared for the feathered beings" given the "superabundance" of food in North America compared with "other countries nearly exhausted by long continued cultivation."[102] I would argue that despite its homespun diction, this passage represents the germ of a full-fledged conservation ethic. Here, Audubon anticipates the core argument advanced by George Perkins Marsh decades later in *Man and Nature*. As Richard Rhodes points out, the calls for preserving wildlife echo throughout Audubon's writing, and heightened with time, as he increasingly sensed the extent of development and the portent of declining game populations:

> In several of his bird biographies Audubon pleads with his readers to spare the lives of species of birds such as Crows and Carolina Par-

> rots that were treated as vermin because they fed on farmers' crops. As he grew older his concern for wildlife conservation increased, less because his perspective changed, as some biographers have speculated, than because he saw the exploitation of the primordial American wilderness accelerating with the accelerating increase of population in the trans-Appalachian West.[103]

Audubon also expressed outrage that commercial exploitation, if left unchecked, threatened even the most abundant species of wildlife, the integrity of habitat, and the indigenous peoples who rely on such resources for subsistence. Traveling in Labrador during the summer of 1833—ten years before the Missouri River expedition—he found the "abundance of fish" in the Bay of Fundy to be "downright astonishing," exceeding even what he had witnessed previously in Florida. Yet he reported that rampant overharvesting already threatened to deplete the resource. He recounted, for example, how a single fleet of "a hundred schooners have brought about 600,000 fish back," and already the very next day would need to seek different waters "because the fish will have grown scarce." To make matters worse, undersized fish were routinely left "to Bears, Wolves, and Ravens on shore" rather than being returned to the water, a "not quite lawful" practice he deemed reprehensible.[104]

For Audubon, Labrador offered yet another example of how even natural abundance of a magnitude unimaginable today was still vulnerable to overharvesting and destruction. He and his companions were astonished by the immensity of flocks of water fowl such as gannets: "I rubbed my eyes, took my spyglass, and in an instant the strangest picture stood before me. They were birds we saw,—a mass of birds of such a size as I never before cast my eyes on. The whole of my party stood astounded and amazed."[105] "The air was now filled with Velvet Ducks," he wrote; "millions of these birds were flying from the northwest toward the southeast."[106] The sight of such a migration inspired awe: "That the Creator should have commanded millions of delicate, diminutive, tender creatures to cross immense spaces of country…is as wonderful as it is beautiful."[107]

Audubon was alarmed by the assault on wildlife he witnessed in Labrador: such extensive "egging, trapping, sealing, fishing, [and] hunting… that populations of furbearers and birds were both already threatened."[108] In his journals he warned of the dire consequences of such unbridled exploitation: "Fur animals are scarce [and] every year diminishes their numbers. The Fur Company may be called the exterminating medium of these wild and almost uninhabitable climes."[109] Audubon described this relentless brutality in "The Eggers of Labrador," an episode published in *Ornithological Biography*, and predicted long-term ecological consequences: "So constant and persevering are their depredations, that these species, which, according to the accounts of the few settlers I saw in the country, were exceedingly abundant twenty years ago, have abandoned their ancient breeding places, and removed much farther north…nature having been exhausted.… This war of extermination cannot last many years more."[110] As Knott concludes, Audubon "saw himself as recording a wilderness that was inexorably receding."[111] In fact, he foresaw that these nurseries on the Gulf of St. Lawrence, if left unprotected, would vanish within fifty years.[112]

At this remote, northern outpost, Audubon had pondered the ethical implications of such unmitigated development, writing in his journal:

> We talked of…the beings best fitted to live and prosper here, not only of our species, but of all species, and also the enormous destruction of everything here…the aborigines themselves melting away before the encroachments of the white man, who looks without pity on the decease of the devoted Indian, from whom he rifles home, food, clothing, and life. For as the Deer, the Caribou, and all other game is killed for the dollar which its skin brings in, the Indian must search in vain over the devastated country for that on which he is accustomed to feed, till, worn out by sorrow, despair, and want, he either goes far from his early haunts to others which in time will be similarly invaded, or he lies on the rocky seashore and dies. We are often told rum kills Indians; I think not; it is oftener the want of food, the loss of hope as he loses sight of all that was once abundant.… Nature herself seems [to be] perishing. Labrador must shortly be depeopled, not only of aboriginal man, but of all else having life, owing to man's cupidity. When no more fish, no more game, no more birds exist on her hills, along her coasts, and in her rivers, then she will be abandoned and deserted like a worn-out field.[113]

Audubon's understanding of how rapidly the historical range of wildlife populations could be reduced by settlement came directly from his grasp of environmental history. He had witnessed the destruction of forests and fisheries and the widespread decline of wildlife, including major species such as the Canada goose and wild turkey and several species extinct today: the passenger pigeon, Carolina parakeet, and presumably the ivory-billed woodpecker. After he had chronicled the shrinking range of particular species, faced with the prospect of absolute extinction, his predictions gradually grew more extreme and his warnings more dire. And the threats to natural abundance posed by development continued to grow toward the end of his career. While his awakening to the ecological perils of development may have happened incrementally with intermittent flashes of insight, there seemed a steady deepening of his appreciation for such issues and a sense of their growing scope.

By the time Audubon arrived on the upper Missouri in 1843, his convictions about preserving wildlife and habitat had been forged by writing for decades about his observations in the field. While journals from the last collecting expedition reveal deepening concerns about accelerating threats to many species, his previous writings had repeatedly addressed these issues as well. As Knott recounts, "Audubon did display increasing concern about the decline of wilderness and of the kind of abundance that was commonplace when he began his wanderings."[114] Moreover, journals recorded on earlier expeditions throughout this career—from Louisville to New Orleans, from Florida to Labrador, long before ascending the Missouri—again and again conveyed narratives of environmental history depicting the dramatic landscape change then under way.

In retrospect, of course, Audubon's concerns were warranted: the widespread decline of wildlife he railed against has indeed come to pass. In the centuries since he lived, the extent of disruption to natural ecosystems has increased relentlessly throughout North America, often with deleterious effects on indigenous species. Audubon recognized threats to environmental integrity and biodiversity long before such concepts were articulated by twentieth-century ecologists. He understood that the range and behavior of birds, including nesting and migration, were subject to change, and that unchecked exploitation could eradicate wildlife. As Streshinsky concludes, "Today Audubon's name is synonymous with the conservation movement. In his time, the woods and skies, the rivers and oceans were filled with birds and animals and fish, [yet] his writing echoes with warnings; he saw the changes civilization wrought, and he was among the first to sound the alarm."[115] Based on his own extensive observations of nature's abundance and destruction throughout North America, Audubon anticipated the broad outlines of conservation in many ways consistent with principles of environmental protection in our time.

Fundamental to these understandings was Audubon's extraordinary sense of ecological memory. Based on observations of the changing range of wildlife populations during his own lifetime, he could extrapolate to consequences for the continent as a whole and recognize what was at stake for future generations. Even today, as ecologists call for contiguous corridors of habitat in the Rocky Mountain West to ensure the survival of threatened wildlife, and as conservationists advocate similar measures in virtually every region of North America he traveled, we follow a trail Audubon blazed.

NOTES

Portions of this chapter are based on conference papers delivered at the Society for the Study of Midwestern Literature (SSML), Midwest Modern Language Association (MMMLA), Great Lakes American Studies Association (GLASA), and the Association for the Study of Literature and the Environment (ASLE).

1. Audubon, *Audubon Reader*, ed. Sanders, 1.
2. Ibid., 9.
3. While there is evidence that some writings attributed to Audubon that were published posthumously may have been altered by subsequent editing, the extent, and intent, of such purported revisions remains open to conjecture, particularly in those cases in which the original manuscripts in Audubon's hand as well as copies made at the time have been lost or destroyed. See the commentary in *The Missouri River Journals of John James Audubon*, edited by Daniel Patterson, for further information.
4. Patterson, *Missouri River Journals*, xvii.
5. Lindsey, *Bicentennial*, xi.
6. Audubon, *Audubon Reader*, ed. Sanders, 8.
7. Lindsey, *Bicentennial*, xi.
8. Streshinsky, *Audubon*, 162.
9. Audubon, quoted in Hart-Davis, *Audubon's Elephant*, 259.
10. Streshinsky, *Audubon*, 343. Quotations from Audubon's writings and the expedition's naturalist, Edward Harris, are presented with their invented spellings and orthographic idiosyncrasies intact.
11. Tyler, "Publication," 139.
12. Hart-Davis, *Audubon's Elephant*, 188.
13. Knott, *Imagining Wild America*, 25.
14. Kohler, *All Creatures*, 7.
15. Audubon, *Audubon Reader*, ed. Sanders, 5.
16. Flannery, *Eternal Frontier*, 6.
17. Tyler, "Publication," 126.
18. Peck, "Audubon and Bachman," 86.
19. Tyler, "Publication," 155.
20. Ewers, *George Catlin*, 502.
21. Ibid., 494.

22. Ibid., 505.
23. Peattie, *Audubon's America*, 275.
24. Durant and Harwood, *On the Road*, 564–65.
25. Audubon, *Audubon: Writings and Drawings*, 725.
26. Tyler, "Publication," 177.
27. Rhodes, *John James Audubon*, 418.
28. Audubon, *Viviparous Quadrupeds*, 9.
29. Ibid., 10.
30. Knott, *Imagining Wild America*, 37–38.
31. McDermott, *Audubon in the West*, 25, 30.
32. Ibid., 31.
33. Ibid., 32.
34. DeVoto, *Across the Wide Missouri*.
35. Audubon, *Viviparous Quadrupeds*, 174.
36. Peck, "Audubon and Bachman," 71.
37. Knott, *Imagining Wild America*, 190.
38. McDermott, *Audubon in the West*, 80.
39. Audubon, *Audubon: Writings and Drawings*, 559.
40. McDermott, *Audubon in the West*, 75.
41. Audubon, *Audubon: Writings and Drawings*, 566–67.
42. McDermott, *Audubon in the West*, 76.
43. Ibid., 85.
44. Ford, *John James Audubon*, 403.
45. McDermott, *Audubon in the West*, 62; Audubon, *Audubon: Writings and Drawings*, 555.
46. McDermott, *Up the Missouri*, 178.
47. Audubon, *Audubon: Writings and Drawings*, 564.
48. McDermott, *Audubon in the West*, 82–84.
49. Ibid., 85.
50. Tyler, "Publication," 154.
51. Peck, "Audubon and Bachman," 76.
52. Buell, *Environmental Imagination*, 415.
53. Rhodes, *John James Audubon*, 87.
54. Audubon, *Missouri River Journals*, 21.
55. Audubon, *Viviparous Quadrupeds*, 180.
56. Audubon, *Audubon and His Journals*, vol. 1, 508–9.
57. Audubon, *Audubon: Writings and Drawings*, 692–93.
58. Streshinsky, *Audubon*, 355.
59. Ibid., 350.
60. Durant and Harwood, *On the Road*, 564.
61. Rhodes, *John James Audubon*, 397.
62. Ibid., 428.
63. Audubon, *Audubon Reader*, ed. Sanders, 196.
64. Durant and Harwood, *On the Road*, 178.
65. Rhodes, *John James Audubon*, 56.
66. Hart-Davis, *Audubon's Elephant*, 184.
67. Audubon, *Audubon Reader*, ed. Sanders, 25–26.

68. Paxton, "Birds," 32.
69. Streshinsky, *Audubon*, 191.
70. Rhodes, *John James Audubon*, 331.
71. Audubon, *Audubon Reader*, ed. Sanders, 68.
72. Ibid.
73. Audubon, *Audubon, by Himself*, 215.
74. Ibid., 220.
75. Audubon notes reverently in *Ornithological Biography* that founding figure Benjamin Franklin favored the turkey over the bald eagle as the country's official symbol.
76. Audubon, *Audubon Reader*, ed. Sanders, 89.
77. Audubon, qtd. in Knott, *Imagining Wild America*, 42–43.
78. Buell, *Environmental Imagination*, 100, 393, 415; Harding and Meyer, *New Thoreau Handbook*.
79. Audubon, qtd. in Jacobs, *Thoreau*, 39.
80. Audubon, *Audubon Reader*, ed. Sanders, 11; Rhodes, *John James Audubon*, 306.
81. Boehme, *John James Audubon in the West*, 35.
82. Audubon, *Audubon Reader*, ed. Sanders, 151.
83. Ibid., 155.
84. Ibid., 151–52, 160
85. Ibid., 156.
86. Audubon, *Audubon, by Himself*, 70.
87. Ibid.
88. Audubon, *Audubon Reader*, ed. Rhodes, 69.
89. Audubon, qtd. in Souder, *Under a Wild Sky*, 154.
90. Durant and Harwood, *On the Road*, 30–31.
91. Audubon, *Audubon Reader*, ed. Sanders, 119.
92. Audubon, *Audubon Reader*, ed. Rhodes, 70.
93. Streshinsky, *Audubon*, 88.
94. Durant and Harwood, *On the Road*, 184.
95. Audubon, *Audubon Reader*, ed. Rhodes, 195.
96. Knott, *Imagining Wild America*, 39.
97. Streshinsky, *Audubon*, 24–25. Occasionally in more reflective moments, Audubon confided that taking specimens for science was not free of regret, such as when he collected a spruce partridge in Labrador: "Her very looks claimed our forbearance and clemency, but the enthusiastic desire to study nature prompted me to destroy her, and she was shot, and her brood secured in a few moments" (Audubon, *Audubon Reader*, ed. Sanders, 218).
98. Knott, *Imagining Wild America*, 40.
99. Durant and Harwood, *On the Road*, 184.
100. Terborgh, "Why We Must," 35.
101. Bent, *Life Histories*, 256.
102. Audubon, *Audubon Reader*, ed. Rhodes, 78.
103. Ibid., 77.
104. Audubon, *Audubon, by Himself*, 229, 230.
105. Audubon, *Audubon Reader*, ed. Sanders, 216.
106. Audubon, qtd. in Hart-Davis, *Audubon's Elephant*, 202.

107. Audubon, *Audubon Reader*, ed. Sanders, 219.
108. Durant and Harwood, *On the Road*, 443.
109. Ibid.
110. Audubon, *Audubon Reader*, ed. Sanders, 82.
111. Knott, *Imagining Wild America*, 190.
112. Rhodes, *John James Audubon*, 383.
113. Audubon's journal, qtd. in Durant and Harwood, *On the Road*, 443.
114. Knott, *Imagining Wild America*, 42.
115. Streshinsky, *Audubon*, xviii–xix.

2

Envisioning Restoration

Gene Stratton-Porter (1863–1924)

I enter a swamp as a sacred place—a sanctum sanctorum.
There is the strength, the marrow, of Nature.

—Henry David Thoreau, "Walking"

What would the world be, once bereft
Of wet and of wildness? Let them be left,
O let them be left, wildness and wet;
Long live the weeds and the wilderness yet.

—Gerard Manley Hopkins, "Inversnaid"

Gene Stratton-Porter's prescient appraisal of midwestern environmental history informed her work as naturalist, photographer, novelist, and conservationist. Her lifelong aesthetic and spiritual appreciation for nature was deepened through a keen awareness of ongoing ecological decline. Born in 1863, she witnessed the escalating pace of landscape change on the edge of settlement and development in Indiana at the end of the nineteenth century. The trajectory of her career progressed from reverence for verdant wetlands to lament for destruction of such vital habitat. The disruption of the region's natural ecological communities awakened in her an environmental consciousness that gave rise to a commitment to stewardship. Single-handedly, she set out to document the immense biodiversity surrounding each of her own two homesteads. In the process, she was moved to advocate for preservation of the region's native species, culminating in the reintroduction of plants on her property known as Wildflower Woods—a monumental undertaking that resembled Aldo Leopold's legendary experiment restoring ravaged farmland along the Sauk River a generation later. While much of her writing seems to constitute an ode to an older order, taken as a whole her life's work anticipated many of the tenets of modern ecology. Still, her contribution to American nature writing and ecological thought—especially the ethical imperative of preserving the biodiversity of species—has yet to receive the level of recognition it deserves.

Biographers such as Judith Long go as far as to compare Stratton-Porter's contribution to conservation to that of her contemporary Theodore Roosevelt, who lauded *her* achievement as an advocate for the preservation and restoration of natural landscapes.[1] She is arguably a seminal figure in midwestern nature writing by virtue of her keen appreciation of environmental history, and is widely credited with influencing subsequent generations of environmentalists, including Rachel Carson.[2] Like American nature writers from Thoreau to Leopold, she advocated conservation through preservation—as well as restoration through the reintroduction of indigenous species. Beyond acknowledging that the natural world provides us with material sustenance, she championed the spiritual benefits of nature, echoing the sensibilities of Transcendentalists she admired, and even ascribing a divine presence to what she deemed wild. In fact, the elevated tone of lyrical passages in her novels and essays alike conveys reverence, even suggesting that she perceived nature as sacred.[3] Her personal creed might well be likened to the following ideals expressed at the close of *Walden*:

> Our village life would stagnate if it were not for unexplored forests and meadows which surround it. We need a tonic of wildness.... At the same time that we are earnest to explore and learn all things, we require that all things be mysterious and unexplorable, that land and sea be infinitely wild, unsurveyed and unfathomed by us because unfathomable. We can never have enough of Nature. We must be refreshed by the sight of inexhaustible vigor.[4]

She believed time in the natural world to be profoundly restorative—a potent remedy for physical illness as well as emotional distress—and hoped that her writing might inspire readers to discover the solace of nature for themselves. Nature, she suggested, might even be likened to an educational institution of the highest order, teaching us moral values. She subscribed to a founding principle of the Nature Study Movement:[5] contact with nature provided balm for the maladies of modern, industrial living. In fact, she incorporated elements from the educational movement into *A Girl of the Limberlost*, while characters in novels such as *The Harvester* attributed virtues such as caution, courage, and patience to childhood experiences in the natural world. Moreover, she sought through her own devotion to nature—like many of her contemporaries during the Progressive Era—to contribute to the ethical and spiritual development of children while re-instilling cherished national values such as independence and self-reliance.

By helping to popularize a proto-ecological perspective, Stratton-Porter's writings contributed to a growing public awareness of environmental history—and served as a precursor to many concerns perpetuated by conservationists to the present day. She hoped that cultivating in her readers an aesthetic appreciation for both the delicacy and grandeur of the natural world could foster environmental sensibilities. Landscapes rich in biodiversity appear frequently in Stratton-Porter's novels as well as in her nature studies, reflecting her growing concern for the preservation of indigenous species. Both her novels and nonfiction nature studies often express reverence for the region's burgeoning wildlife, especially birds and moths indigenous to the wetlands and forests of central Indiana. Beyond devoting herself to documenting the abundance of the region's remaining untamed landscapes, she bore witness to the unremitting destruction of wild places that she knew intimately. Shadowing the saga of natural abundance was a narrative of environmental decline, as woodlands were logged off, wetlands dredged and drained, oil wells drilled and pumped dry, a process recounted in novels such as *Freckles* and *A Girl of the Limberlost*. She condemned development that supplanted entire ecosystems; extensive wetlands were rapidly being lost throughout the Midwest, including Limberlost Swamp. As her late writings attest, she arrived at deep convictions about propagating endangered plants after seeing tract after tract of woodland cleared for grazing and cultivation at the expense of native species.

Stratton-Porter's interest in preserving indigenous species grew out of her extensive experience exploring Indiana's natural bounty, as well as witnessing its ongoing demise. In fact, her convictions about stewardship arose much like those of earlier nineteenth-century naturalists such as John James Audubon, who had also arrived at conservation ideals while lamenting the decline of wildlife due to exploitation and habitat loss. She recognized the importance of protecting the integrity of preexisting ecosystems, especially verdant wetlands. It was not until 1922 that she published her clarion call to environmental advocacy, "All Together, Heave," a fitting culmination to her career as a naturalist. In a sense, this single essay distilled the entire arc of her development as a conservationist: from initial wonder at the beauty, grandeur, and abundance of nature, to profound appreciation for environmental history, accompanied by poignant laments over ecological degradation. She recounted the recurring saga of natural abundance squandered, both locally and across the continent:

> During my own childhood, I recall the ascending smoke of countless burning log heaps... golden oak, walnut, red cherry, bird's-eye maple and hickory trees... indiscriminately stacked and burned where they fell in order that the land upon which they stood might be used for the cultivation of crops... untold billions that went up in the smoke of the most wanton waste this world ever has known. The animal life receded before the destruction of the forests, the buffalo ranging in countless herds are now practically extinct.... The wild pigeon that once broke down branches of forest trees with the weight of their numbers, now gone forever.[6]

This litany of ongoing loss was echoed perhaps even more forcefully in her *Tales You Won't Believe*, published posthumously in 1925: "In pain, confusion and futile protest I watched the buffalo disappear from the prairies; the red horse, the bass, and the wood ducks from the forest; while as exquisite wild flowers as the circle of the globe ever knew vanished from the face of the earth."[7] In her estimation, stemming the tide of destruction—and the prospect of extinction—had become an ethical imperative.

Stratton-Porter's convictions about conservation were a lifetime in the making. Beginning with her Hoosier childhood, her appreciation for nature was firmly rooted in environmental history, particularly what was at stake when the integrity of entire ecosystems was being so rapidly degraded by development. When she was born on her family's farmstead outside Lagro, Indiana, the original forests along that stretch of the Wabash watershed had already been felled and converted to cropland, orchards, and pasture. Yet as biographer Bertrand Richards recounts, "It was land removed only a step from wilderness. On the prairies to the west, buffalo still abounded. The rivers and streams were filled with fish. Birds were everywhere. The great flocks of passenger pigeons still flew over Indiana. Wild turkeys, prairie chicken, and quail were hunted for the table. Deer and bear were plentiful, as were smaller game animals."[8] She later characterized the area adjoining an extensive stand of virgin forest as being exceedingly beautiful and particularly valuable as habitat for nesting birds. An essay introducing later editions of *At the Foot of the Rainbow* depicted the Stratton family homestead, Hopewell Farm, where her insatiable and lifelong passion for exploring the natural world was instilled. The family grew and preserved their own food, tending small orchards of apples and peaches, and allowing some fields to lie fallow while still producing enough corn to feed both themselves and their livestock, including cattle, chickens, geese, horses, pigs, and turkeys. The soil proved particularly fertile, she believed, because "the land was new, a large part of it having been cleared and put into cultivation by [her] father."[9] In an era of such self-sufficient farming, agrarian families readily developed an aesthetic appreciation for and an ethical bond with the land: a respect for the soil and a vested interest in stewardship.

In the latter half of the nineteenth century, small family farms of the Midwest shaped the hybrid landscape of Stratton-Porter's youth, a matrix of wild and cultivated spaces inadvertently providing a great deal of habitat, as historian William Leach observes: "While they sacrificed virgin forests and ecosystems . . . ponds, woodlots, hedgerows, stone walls, open fields along roadsides, and meadows by stream or riverbeds for grazing cattle . . . also serv[ed] as likely habitats and hideouts for animals," a verdant environment also teeming with birds, butterflies, and native plant species.[10] When she returned as an adult, however, hoping to reacquire the family's old farmstead, she was alarmed to find the place irrevocably changed. As Mary Obuchowski recounts, "Later she visited Hopewell Farm, her childhood home, and found the creeks and streams gone, the land altered almost beyond recognition."[11] The house and surrounding trees were also gone, wetlands dredged and drained. "The hills were short and ploughed down, filling and obliterating creeks and springs," she reported, while "most of the forest had been cut, and stood in corn."[12] The conversion to monocultural cropland was so complete, she realized, that it would never again be restored to the lush landscape that she had cherished as a child.

Stratton-Porter recognized that many readers at the time mourned the passing of agrarian ways of life as she did, and had begun to grow nostalgic for what must have seemed a golden age: tranquil landscapes of their youth when local ecosystems were only first beginning to be compromised. She sensed a gradual shift in her audience that she characterized as a deepening desire for stories steeped in the verdant natural world that they too remembered.[13] Perhaps she helped to cultivate the broader audience she so keenly hoped to reach: a readership in many cases a single generation removed from the nineteenth-century agrarian existence of subsistence family farming, with whole communities acutely attuned to the progression of seasons. Her family's move from Hopewell Farm to the town of Wabash in 1873 when she was eleven seems emblematic of this period of increasing industrialization, when populations were shifting dramatically from rural to urban areas, a process exacerbated a generation later by increasingly large agricultural enterprises and the exodus of small farmers.

While literary critics of Stratton-Porter's day typically dismissed her fiction as sentimental romance, she became one of the most widely read American authors of her generation. Her many novels succeeded in charming millions of readers; in fact, five of them sold over a million copies apiece.[14] She lived to see several titles, including *The Harvester*, published throughout Europe, including England, France, Germany, and Scandinavia, as well as a production of the film *Freckles*, released by Paramount in 1917—the first of two dozen movies based on her books. Such popularity broadened her influence by introducing generations of young readers to a proto-ecological perspective. In fact, *A Girl of the Limberlost* (1909) and *The Harvester* (1911), among her best-known works, draw from her own experience to invoke regional environmental history and themes related to the preservation of nature. *The Harvester* sold a quarter of a million copies in the first year alone and led admirers to seek out her previous books, particularly *A Girl of the Limberlost*.[15] Readers were thrust into the flux of landscape change—and the soul searching of those who revered the verdant wetlands while watching their biodiversity being eroded before their eyes.

The popularity of Stratton-Porter's novels completely eclipsed her other, more lavishly illustrated works, such as *Music of the Wild*, devoted wholly to extolling the wonders of the natural world. So she negotiated publishing contracts enabling her to alternate between lucrative romance novels such as *A Girl of the Limberlost* and *The Harvester* and the more costly format she actually preferred: volumes of prose and photographs addressing natural history. Although the "nature books" sold far fewer copies than her popular fiction, she accepted the financial sacrifice. Her editors initially underestimated the extent of public interest in the natural world, reportedly asking her to remove extended depictions of nature from early novels. Once her books had propelled her to literary stardom, however, they encouraged her to incorporate more of them.

Stratton-Porter believed the scientific material in her novels to be unimpeachable in its accuracy, being grounded in her formidable experience collecting specimens and photographing in the field. Moreover, she felt a moral obligation *as a novelist* to appeal to the higher nature in her readers and instill sensibilities that she deeply admired such as altruism and stewardship of the natural world. Moreover, "by integrating clear and factual natural descriptions into fiction rather than using romanticized images," as eco-feminist Cheryl Birkelo asserts, she "spreads ecological understanding of life's miracles to her audience while still acknowledging the dilemma of conservation and preservationist ethics and the relationship between humans and their environments."[16] By "attempting to inculcate sympathy with nature," as environmental historian David Armitage points out, she "tied her work to the moral imperative of conservation."[17] As Birkelo concludes, while "critics have overlooked the environmental import of Stratton-Porter's principal works of fiction . . . she infuses them with the same message of conservation and environmental concerns that she addresses in her nonfiction . . . [especially] responsible stewardship of the land."[18] Both genres expressed wonder: "Appreciation of nature underlies the plots in Porter's fiction," as Obuchowski observes, adding that "her non-fiction books and nature articles also embody reverence for ancient forests and delight in the beauty of flowers, birds, and moths."[19] The novels so often served as a vehicle for addressing conservation that they were without question of a piece with her nature studies. Consequently, only when the fiction and nonfiction are considered together—and her oeuvre viewed in its entirety—can her overall contribution to midwestern environmental history be fully appreciated.

The aim of Stratton-Porter's nonfiction was above all to document the many species of moths and butterflies, as well as birds that nested in the wetlands, reveling in what we today call species biodiversity. In the 1916 essay "My Work and My Critics," Stratton-Porter explained that she "went afield and brought in the first-hand material for five nature books, illustrated them with half-tone from life, and water-color paintings; [and] wrote stories containing thousands of natural history references."[20] In volumes of nonfiction such as *Homing with the Birds* published in 1919, for instance, she implored her readers to rally to the cause of conservation. Her passionate pleas on behalf of preserving nature expressed in late essays such as "Shall We Save Natural Beauty?" were rooted in a lifetime of close observation in the field. In fact, Stratton-Porter's enduring concerns, such as "her appreciation of nature, her awe at its creator, [and] her desire to infuse readers with her enthusiasm for it...[dominate] her poetry as well."[21]

In many respects, Stratton-Porter's career as an author, artist, and naturalist bridged the sciences and humanities, biology and literature, aesthetics and ethics. Indeed, the range of her achievement might be likened to that of earlier naturalists whom she admired and emulated, above all Thoreau.[22] Like Thoreau, Stratton-Porter's reflections on natural history and philosophy were grounded in intimate acquaintance with place. Living for nearly two decades, between 1895 and 1914, on the edge of Limberlost Swamp—once a verdant, thirteen-thousand-acre wetland—she observed a microcosm of midwestern landscape change, the swift and seemingly irreparable demise of natural ecosystems denounced by the region's literary naturalists to this day. When she first saw Limberlost Swamp, it was largely unadulterated: "When I arrived," she would later write, "there were miles of unbroken forest...it was practically a tropical jungle."[23] "Only a few trees had been taken out; oilmen were just invading it.... It had not been shorn, branded, and tamed."[24] Because it was still relatively undeveloped, the swamp was filled with an extraordinarily rich mix of indigenous species, the very biodiversity that inspired her nature studies. On arrival, she characterized this remote section of the Limberlost as "original timber," a forest composed of a diversity of trees such as ash, chestnut, hackberry, and linden as well as at least three species of oak. Yet the idyllic quality of even such an isolated area was not to last. While much of the forest was being logged off and extensive tracts plowed to plant wheat and corn, she celebrated pockets of native flora tucked into the corners of the traditional split-rail "snake" fences, including fruits such as raspberry, gooseberry, and wild grape, as well as an assortment of native trees and a host of wildflowers. She reveled in the abundance of wildlife sustained by such microhabitats. The profusion of native plant species found in such islands of biodiversity—alder, aster, bittersweet, elm, goldenrod, honeysuckle, maple, milkweed, roses, saffron, sumac, teasel, and thistle—created thickets that provided sanctuary for birds that had become scarce elsewhere due to the destruction of habitat suitable for nesting and feeding. The old "snake-fences" that she exalted as tiny refugia were "weighted with loads of growing shrubs and vines, and so on, and on and under them home field mice, moles, rabbits, chipmunks, lizards, birds of low habit, night moths, and bugs and insects of innumerable species."[25]

Stratton-Porter discovered that the most remote and undeveloped sections of the swamp continued to support a greater variety and profusion of birds. Development of the area proved unremitting, however, precipitating profound landscape change. We can only imagine the depth of her despair as the wetlands that had inspired such reverence were rapidly leveled before her eyes.[26] Her five collections of nature essays function as ecological time capsules: "recollections of a time and world that no longer exist and cannot be reclaimed…with the sense of loss inherent in remembered beauty."[27] She lamented the "dismantling" of the swamp, portraying the destruction in popular novels such as *Freckles* and *A Girl of the Limberlost*, as well as in corresponding nonfiction, particularly *Moths of the Limberlost*. The latter work, perhaps more than any other, established her reputation as a naturalist and demonstrated the depth of her scientific knowledge, proudly grounded in her extensive observations in the field.

Stratton-Porter railed against the seemingly relentless process by which natural landscapes were effectively erased. She denounced the "devastation" of wetlands as hardwoods were harvested, ditches dredged, oil extracted, and large-scale farms established. Once trees were logged off, brush was often burned to make way for grazing livestock. "The cleared fields come next, and as they march with inexorable force, they push back the woods farther and yet farther."[28] She was appalled by the demise of the verdant wetlands of the Limberlost, a process recounted in *Homing with the Birds:* "Five years ago through the work of farmers and lumbermen, my immediate territory had been cleared, drained, and put under cultivation, until the birds had flown, [and] the flowers and moths were exterminated."[29] Expressing ethical sentiments fervently in terms of conscience, she declared that such "wanton destruction is one of the blackest of sins."[30] Moreover, she cautioned that landscape change had become so rampant and dramatic as to be "changing the forces of nature" to the point of climate change—much as George Perkins Marsh and Alexander von Humboldt had predicted by the middle of the nineteenth century the peril of desertification worldwide. "If men in their greed cut forests that preserve and distill moisture, clear fields, take the shelter of trees from creeks and rivers until they evaporate, and drain the water from swamps so they can be cleared and cultivated," she warned the water cycle itself would be fundamentally disrupted. Echoing Thoreau, she declared that dredging and draining wetlands while felling forests was tantamount to cutting down the clouds.[31]

Descriptions of ongoing environmental decline appear frequently in her novels as well. *A Girl of the Limberlost* depicts the dismantling of the swamp that had inspired her finest works of natural history. The impact of such development was cumulative, she realized; the decline of species that had historically thrived sometimes culminated in the obliteration of entire ecosystems. Both preceding and during her lifetime, many kinds of game had been hunted out completely in the Limberlost Swamp, including predators such as bobcats, bears, and wolves, and their prey, above all beaver. Even whitetail deer were gone and would not be reintroduced to the region until generations later. Extrapolating from environmental destruction that she had observed, she cautioned that unchecked development could have dire consequences, including the widespread extirpation of wildlife and conceivably the extinction of some species. While such a concern is distressingly commonplace today, it was a relatively novel proposition when she was writing a century ago.[32]

Stratton-Porter lived and worked, after all, generations before modern ecology was founded as a scientific discipline. Being largely self-taught, she was in some senses a holdover from an earlier era of serious amateur naturalists, though she did not have the advantages of privilege or wealth that enabled many of her predecessors, such as Darwin and von Humboldt, to secure the requisite education, leisure, and means to pursue fieldwork near and far. In the first decades of the twentieth century, collecting specimens such as butterflies and moths in the Victorian manner was still a remarkably popular pastime, a practice rooted in the nineteenth-century traditions of natural history.[33] She published articles, such as "Hidden Treasures" in 1912, that provided detailed, technical directions for readers interested in properly preserving specimens for their own collections. With its emphasis on identification and classification based on taxonomies of species, collecting had surprisingly widespread appeal. In an age of colonialism, Americans packed curio cabinets with fossils, specimens, and artifacts that might have been gathered from distant continents and featured species from exotic landscapes such as the tropics.

Stratton-Porter's sensibilities regarding conservation were neither simple-minded nor nostalgic, but rather reflected a clear-eyed pragmatism. She recognized how wildlife was essential to the workings of natural systems as a whole, and ultimately vital to the survival and well-being of humanity. She reasoned that agriculture, particularly as it was practiced at that time, remained reliant on the integrity of surrounding wild ecosystems. Before the introduction of toxic pesticides, conventional wisdom held that birds that kept insects in check were essential for successful farming. In order to "insure magnificent crops of large and small fruits," she declared, population declines would have to be reversed: "protect and increase birds as to raise their numbers again to such flocks as I knew in childhood."[34] To achieve such ends, however, she opposed the predator control measures prevalent at the time. Exterminating raptors, crows, and jays that preyed on songbirds and their nests was not a prudent policy, in her estimation, since it would inevitably disrupt the overall natural balance of wildlife. This view anticipated enlightened wildlife management practices that pioneering midwestern ecologists such as Aldo Leopold and Paul Errington would not arrive at until decades later.

Since birds had become differentiated through natural selection, Stratton-Porter contended, both avian anatomy and behavior had been profoundly shaped by various types of habitat, leaving traces of relationship to place. Daughter of a Methodist Episcopal minister, she seemingly reconciled Christian doctrine with evolutionary theory by viewing nature as divine—including the process of natural selection itself.[35] She interpreted the Biblical account of creation in seven days as largely symbolic, positing instead "all the time necessary for the evolution of the earth and development of plant and animal life."[36] She cited the emerging fossil record, particularly the celebrated archaeopteryx, which linked birds to reptilian ancestors during the age of dinosaurs, as in keeping with a Darwinian theory of the evolution of species. Writing in 1909, she took for granted an account that was shared by the scientific community of her time. "Animal life," she concluded, "originated in the water, and developed there for untold time before it appeared upon the land; and with different environment took on different forms."[37] In light of Darwin, she recognized that over time species had in fact *evolved together*.[38] Moreover, she appreciated that instincts for behavior such as migration had developed through interaction with the changing environment over eons. She particularly marveled at the range of migrations that spanned nations and even continents, noting that the precise timing required to survive such flights demonstrated what she referred to as their "unerring knowledge of workings of nature."[39] Though lacking formal training and academic credentials, Stratton-Porter challenged the claims of scientific authorities of the day whose acquaintance with behavior and even habitat was vague since their examinations relied largely on preserved specimens. The theories of such "closet" practitioners and "cabinet" naturalists lacked the veracity that came from firsthand experience in the field. In addition, given her extensive observations in the wild, she was perfectly positioned—as Audubon had been nearly a century earlier—to recognize such deficiencies and correct errors of fact and question assertions by members of the scientific establishment she knew to be based on conjecture. In Stratton-Porter's time, after all, a highly observant citizen-naturalist could still make original contributions to the study of natural history. The biological sciences had only recently turned in the direction of the kind of descriptive life studies depicting behavior and relationships among species that would later become the centerpiece of modern ecological thought.

Such intensive fieldwork demanded unfettered access to places still largely unadulterated by development. It was there that she collected specimens and conducted the work of a consummate amateur naturalist. Following the loss of her beloved swamp and disheartened by the utter destruction of the Limberlost, she felt compelled to make a new home elsewhere.[40] "Now it has so completely fallen prey to commercialism through the devastation of lumbermen, oilmen, and farmers," she lamented, "that I have been forced to move my working territory and build a new cabin about seventy miles north...when my working grounds were cleared, drained and ploughed up, literally wiped from the face of the earth."[41] Once the abundance of the wetlands that had inspired her were so thoroughly degraded, she chose to relocate to "Limberlost North" outside Rome City in northeastern Indiana. The place had tremendous personal resonance: she had met her future husband there when attending summer Chautauquas as a girl in the 1880s while her father lectured. She purchased fourteen wooded acres on the lake and designed a palatial log-sided house overlooking an unsettled one-hundred-acre island. At the main entrance facing the lake, on a banister at the foot of the stair, she placed an early electric lantern shaped like a flame to remind her of childhood nights watching from her family's homestead at Lagro, Indiana, on the Wabash River as Indians passed by torchlight. She wrote approvingly of early experiences interacting with several Native American peoples, including the Meshingmesas who lived near her family's homestead at Hopewell Farm: "All my life I have been steeped in Indian tradition and history, having been born and reared near the camping ground of the Potawotamies and the Miamis.... When I lived in Wabash, Indiana, I was bounded on the west by the Miami Reservation."[42] Her popular novel *Laddie*, published in 1913—at the time she was overseeing design and construction of the cabin at Limberlost North on Sylvan Lake—includes detailed recollections of Hopewell Farm where her fascination with natural history had been first awakened. Perhaps she wished her new home to rekindle the exaltation in nature she had experienced in her youth. Even before preparing the building site for a home, she sought to establish premium habitat for wildlife, above all birds. While deadfall and rotting logs were removed, many native species of trees growing on the wooded acres, including ash, beech, chestnut, elm, hackberry, linden, maple, oak, and tulip poplar, were fastidiously tended. Above all, the place on Sylvan Lake provided a safe haven to escape the last dismantling of Limberlost Swamp and an opportunity to preserve native species.

The largely un-despoiled agrarian landscape of Limberlost North on Sylvan Lake did indeed seem to reawaken and perhaps even deepened Stratton-Porter's connection with nature. She described the surrounding area as being largely unadulterated at that time. In "A Little Story of Her Life and Work," a biographical essay introducing the 1916 edition of *At the Foot of the Rainbow*, she reveled in the verdant local wetlands: "many lakes, miles of unbroken marsh, and a far greater wealth of plant and animal life than existed during my time in the southern part.... In one season I have located here almost every flower named in the botanies as native to these regions."[43] The secluded dwelling proved to be an ideal setting for writing a number of her later books, both novels and nonfiction nature studies such as *Friends with Feathers* and *Homing with the Birds*. While dividing her time between California and Limberlost North in the final years of her life, she also authored *The Fire Bird*, a volume of poetry depicting Native Americans in a largely romanticized manner reminiscent of Henry Wadsworth Longfellow's 1855 epic poem *The Song of Hiawatha*, replete with an Indian princess and medicine men, mysterious potions and dances of enchantment. *The Fire Bird* was purportedly vetted by photographer Edward Curtis (to whom the book is dedicated), whose own monumental twenty-volume *The North American Indian* was still under production, as well as by famed western painter and sculptor Charles Russell.[44]

Stratton-Porter championed the photographic approach to illustrating natural histories as opposed to the customary practice of collecting dead specimens for subsequent examination, which perpetuated an emphasis on anatomy and classification completely devoid of ecological context. Prior to the onset of photography, of course, naturalists had relied on paintings and drawings—often by their own hand—to depict species, whether rendered from specimens or observed in the wild. Audubon's experience a century earlier is emblematic in that it spans several generations of printing technology. His journey to London and Scotland to secure a publisher for *Birds of America* is the stuff of legend. There he recruited a renowned printer to make engravings of his original oil paintings on copper plates. The life-sized prints were then hand tinted to the most exacting standards. Not until the end of Audubon's life, at the middle of the nineteenth century, did the ascent of lithographic printing enable the production of a smaller, affordable octavo edition of the epic ornithological portfolio.

Though Stratton-Porter was an accomplished artist, she became one of the pioneers of wildlife *photography*.[45] Emerging photographic technology, she quickly recognized, provided an unprecedented tool for biologists to identify species and document behavior without sacrificing the lives of so many specimens in the process.[46] In fact, she took nineteenth-century naturalists such as Audubon to task for what she viewed as excessive collection of specimens, resulting in unnecessary slaughter in the wild. "I wish to register my most serious protest against the work of professional ornithologists," she declared, "in hastening the extermination of rare and unusual birds in their greed to secure specimens for museum collection.... Let any bird be reported as rare; immediately pursuit of it with gun begins.... The rare bird once more a fugitive at the hands of science! This is a real crime, and a law should be enacted to punish it." She advocated legislation that mandated moderation on the part of scientists and regulated sport and market hunting.[47]

Stratton-Porter gained notoriety traipsing through Limberlost Swamp with cumbersome, large-format cameras, work made treacherous by lugging such unwieldy equipment. Defying strict gender prescriptions of the time while experimenting with novel photographic technology undoubtedly contributed to her image as daring and iconoclastic. The scientific establishment of her day questioned the credibility of a self-taught naturalist, especially in an era before women rose to the fore in such disciplines.[48] Yet as she proceeded to publish arresting images such as intimate portraits of birds nesting in the wild, her reputation as a pioneering nature photographer grew. The lavishly illustrated *Moths of the Limberlost*, for example, was intended to be an exhaustive account of species native to the Indiana wetland and their life cycles. Life studies of each species were accompanied by illustrations painstakingly prepared by applying watercolor pigments to her own photographs. Remarkably, the scope and approach of descriptions published in *Moths* over a century ago in many respects prefigure the content and presentation of corresponding entries in the standard field guides of our time.

According to her daughter Jeannette Porter Meehan's account in *The Lady of the Limberlost*, a compendium of reminiscence and correspondence, Stratton-Porter's concerted study of moths had begun with the casual collection of cocoons while photographing birds. She nurtured the cocoons in her home until the moths hatched—going so far as to spray them to simulate rain. Eventually, she went to terrific lengths to witness the entire life cycle of both moths and butterflies, raising caterpillars from eggs as well as observing their metamorphosis upon emerging from cocoon or chrysalis. By working with living moths—rather than the preserved specimens typically relied on by scientists of the time—she represented colors with unprecedented precision while also capturing the moths' natural movements. Indeed, she took extraordinary measures to depict the delicate gradations of pigmentation on the wings of a precious specimen, such as the rare Yellow Emperor moth: "shades of lavender running into purple... liberally decorated with lavender...living moths were beautiful beyond description."[49] To achieve absolute verisimilitude, she meticulously tinted each black-and-white print by hand—a painstaking and exacting process.[50] She relished rendering moths freshly emerged from cocoons while their delicate pigmentation was still most vivid. In her estimation, publishers had only recently devised printing techniques sufficiently sophisticated to reproduce such subtle variations in coloration accurately.

Critics generations later mistakenly assumed that she had indiscriminately collected specimens for sale. In fact, reflecting respect for the ethics of specimen collection, she customarily released every moth after taking photographs, preserving just a single pair—a male and a female for each species—as a reference when preparing hand-tinted prints as illustrations for her books. "All I have ever mounted I used as models for painting and descriptive copy," she assured readers. "I am sure I never took more than I really needed in order to do my work well."[51] She directed her readers interested in collecting accordingly: "The nature lover can study a moth and set it free." And for more serious amateur naturalists, especially those intending to make their own illustrations, she advised keeping only "a perfect pair of each species for reference. There is no necessity for taking large numbers, and ruthless destruction."[52] She considered the process of illustration warranted only if rare species were procured with the utmost discretion and in service of art, science, and ultimately education.

The resulting photos and illustrations were clearly intended to express wonder and awe and presumably to cultivate similar sensibilities in her readers. By sheer grit and ingenuity, she produced the illustrations to grace both her novels and works of natural history. Her arresting images of birds photographed in the wild, much like those in Audubon's *Birds of America*, were often animated and imbued with purpose, reflecting what she termed *joy*, *grief*, *pain*, *fear*, and *greed*, as well as *anger*, *curiosity*, *love*, and *pride*, though she conceded that "exactly how *much* they know, and exactly *how* they know it we can only surmise."[53] An aesthetic appreciation, she understood, could not only become the basis for a personal bond with the natural world but kindle an interest in the ethics of environmental preservation. Consider the visual imagery used by those advocating conservation to the present day to engage the public in support for their causes and campaigns—above all, striking photographs of seemingly pristine landscapes and iconic wildlife. "The appreciation of natural beauty" in Stratton-Porter's time, as Armitage observes, "propelled nature study conservation: it promoted science while simultaneously engaging the romantic longings of turn-of-the-century nature lovers."[54] She was arguably among the first to grasp how effectively photography could promote conservation. Reveling in nature's beauty and grandeur, she reasoned, was just a short step away from decrying its demise or rallying to its defense.

Similarly, Stratton-Porter conceived of her widely read fiction as a vehicle for spreading her beliefs about protecting nature through stewardship. She undoubtedly wished that her books might ultimately inspire readers to explore nature for themselves. In fact, in *Moths of the Limberlost*, she expressed this very aim: to acquaint "people who could not go afield to see for themselves, and to teach those who could to protect helpless, harmless things."[55] Characters in several of her novels model an attentive and reverent stance toward nature much like her own. Among the most inspiring passages are those portraying characters responding reverently to the natural world, as if imbued with the author's own passions.

The overall premise of *A Girl of the Limberlost* clearly derives from personal experience. Two characters in the novel seemingly combine to embody Stratton-Porter's own sensibilities: the book's protagonist, Elnora Comstock, and her delightfully eccentric mentor, referred to only as the "Bird Woman" (incidentally a nickname of the author herself), who was Elnora's primary customer for specimens of butterflies and moths.[56] Elnora is a country girl who sought to advance her education by entering high school in a neighboring community. But to afford tuition and books, she considers selling off rights to develop the three hundred acres adjacent to a swamp that she is soon to inherit, in effect recapitulating Stratton-Porter's own purchase of a remote 320-acre parcel as an investment. One could profit handsomely when timber was logged off and oil wells drilled before the parcel itself was resold for agricultural use, as happened with the dismemberment of the Limberlost. Yet like the author herself, Elnora is troubled by the demise of the verdant wetlands, stripped of cultural as well as ecological treasures. The plundering and depletion of ancient Native American artifacts seemingly keeps pace with the felling of the forest: "they are clearing the swamp so fast. Every year it grows more difficult to find things, and Indian stuff becomes scarcer."[57] Increasing scarcity of biological specimens could be roughly quantified: following just three years of intensive timber cutting, for instance, the yield of rare moths from a day's collecting had fallen twelvefold.[58] The growing rarity of many species of moths made them all the more precious not only to collectors but as evidence of the integrity of wetland ecosystems as a whole. In fact, the novel portrays ongoing demolition of the Limberlost—ultimately leading Elnora to declare, "the swamp is almost ruined now"[59]—and suggests the dire consequences of such unbridled development:

> Men all around were clearing available land. The trees fell whenever corn would grow. The swamp was broken by several gravel roads, dotted in places around the edge with little frame houses and the machinery of oil wells.... Whenever the trees fell the moisture dried, the creeks ceased to flow, the river ran low, and at times the bed was dry. With unbroken sweep the winds of the west came, gathering force with every mile and howling and raving, threatening to tear the shingles from the roof, blowing the surface from the soil in clouds of fine dust, and rapidly changing everything.[60]

Published two decades before the monumental erosion of topsoil during the Dust Bowl, this passage seems hauntingly prophetic.

Throughout the novel, issues of social justice are in play involving rural poverty and economic opportunity. A neighbor urges Elnora's mother to embrace extractive activities such as logging and oil drilling as well as cattle ranching—a microcosm of landscape change rampant in the Midwest at the time:

> "Any day you say the word you can sell six thousand worth of rare timber off this place easy. I'll see to clearing and working the

> field.... I'll buy you more cattle to fatten. All you've got to do is sign a lease to pull thousands from the ground in oil, as the rest of us are doing all around you."

But she vehemently refuses, choosing instead to honor the legacy of her late husband:

> "Cut down Robert's Trees!" shrieked Mrs. Comstock. "Tear up his land! Cover everything with horrid, greasy oil! I'll die first."[61]

Mrs. Comstock fervently defends her decision to forgo the financial rewards of allowing development of her land adjoining the Limberlost Swamp. Moreover, she expresses her desire to carry on the tradition of pioneering settlers while implicitly preserving the integrity of the surrounding wild environment:

> In all the present-day rush to be modern, I have preferred to remain as I began. My husband and I took up this land, and enough trees to build the cabin, stable, and outbuilding are all we ever cut.... I hear considerable about the value of the land, the trees which are on it, and the oil which is supposed to be under it, but as yet I haven't brought myself to change anything. So we stand for one of the few remaining homes of first settlers in this region.[62]

Secondly, a feminist thread emerges as Elnora seeks access to higher education in the era of women's suffrage by striving to save for college while still in high school. To afford tuition and books, she sells treasures gleaned from the wetlands such as specimens of rare moths and butterflies, as well as Indian artifacts—emblems of an ecological and cultural legacy already jeopardized and therefore to be all the more cherished. Some critics have mistakenly charged that the collecting of specimens for sale depicted in the novel represents an anti-environmental message: the exploitation of nature for personal gain. In the story, however, such collecting clearly has a redeeming purpose—primarily to illustrate nature books, much as Stratton-Porter herself did, raising moths from cocoons and releasing them after making illustrations intended to expose a wide reading audience to the exquisite beauty of such fragile creatures.

In *A Girl of the Limberlost*, Elnora seemingly personifies the Nature Study Movement of Stratton-Porter's time. Her mentor, a character referred to as the "Bird Woman," has alerted her to the use of natural specimens in elementary schools, explaining that "grade rooms want leaves, grasses, birds' nests and cocoons."[63] As an amateur naturalist skilled at collecting specimens in the wild, Elnora can readily supply that demand. She soon proves to be a savvy entrepreneur, marketing such materials to the schools, vowing to "collect specimens of all tree leaves before they fell, gather nests and cocoons…anything the grades could use. She would see the superintendent that night about selling specimens to the ward buildings."[64] Gradually Elnora begins to assume a direct instructional role with the schoolchildren, identifying specimens for teachers while explaining to students their significance in relation to the wetland niches where they had been collected.

Rather than furthering her education by attending college immediately after high school, Elnora instead accepts a position as lecturer in Natural History charged with infusing grade school curriculum with biological knowledge grounded in the local landscape.[65] The job description for Elnora's lectureship details the following responsibilities: "exhibiting and explaining specimens of the most prominent objects of nature: animals, birds, insects, flowers, vines, shrubs, bushes and trees…appropriate to the seasons and the comprehension of the grades," precisely the kind of educational initiatives advocated by Anna Comstock and the Nature Study Movement.[66]

Comstock's seminal pedagogical text *Handbook of Nature Study*, first published in 1911, championed the teaching of natural history.[67] Like Stratton-Porter herself, Comstock illustrated prominent volumes of natural history, particularly entomological works. In fact, her influential guide to introducing biological sciences in the schools espoused an approach and philosophy closely resembling those that represented in several of Stratton-Porter's novels and nonfiction nature studies such as *Moths of the Limberlost* and *Homing with Birds*. According to its proponents, the movement was to counter the ongoing environmental decline that Comstock characterized as having "exterminated many species of birds, destroyed many interesting and harmless wild animals, hacked down our trees ruthlessly and cleared our streams of valuable fish."[68] The nation had only begun to take stock of landscape change and habitat loss over the last several centuries. Comstock's personification of a science educator might well have provided a model for introducing natural history in the schools as portrayed in *A Girl of the Limberlost*.[69]

The Nature Study Movement in American schools expressly encouraged the cultivation of delight: celebrating wonder while acknowledging a spiritual dimension to the appreciation of nature. Writing in 1912, Stratton-Porter herself characterized the growing popularity of the approach and its positive impact on pupils: "Schools all over the country are now teaching every branch of nature work. . . . Many teachers have told me that they cannot arouse the unbounded enthusiasm, and awaken in children interest in and love for the outdoors, so effectively in any other way, as they can by bringing cocoons into the schoolroom and allowing pupils to watch these wonderfully-colored, big, bird-like moths emerge, dry, and develop their wings, and take flight."[70] Witnessing the mysteries of metamorphosis, she believed, inevitably fostered ethical sensibilities and promoted greater responsibility, since "the more you know about the intricacies of moth development, the farther you trace the slow evolution that has resulted in this complicated form of life, the more carefully you will handle them, the more respect you will have for them."[71]

Several premises of the Nature Study Movement anticipated the tenets of ecological thought articulated by wildlife biologists such as Aldo Leopold a generation later, including an emphasis on direct experiential knowledge of nature and place and the ethical imperative of preservation. What began as aesthetic appreciation for natural beauty, it was thought, could readily catalyze moral convictions about protecting the environment. As David Armitage contends, "it is ecology that translates Progressive Era nature study and conservationist concerns into modern environmentalism."[72] Moreover, the Nature Study Movement can be viewed as the "foundation of environmental education," as well as "a forerunner of environmentalist ideologies such as deep ecology."[73] As early as 1907, nature study had been widely introduced in schools across the nation; by 1914, twenty-three states had officially endorsed its place in the curriculum. The movement continued to gain in popularity, spreading its influence to virtually every state in the nation by 1920.[74] Even while incorporating aspects of biological science, such natural history curricula reflected a deep neo-Romantic distrust of ever more mechanized modernity, and skepticism about "progress" being contingent on technologies that separate us from nature. Similarly, the School Arts Movement of the period, championed by Henry Turner Bailey, advocated the cultivation of aesthetic appreciation by close observation of nature; extensive classroom use of specimens such as flowers, grasses, leaves, and seeds; and an art curriculum organized on the basis of nature's cycles and seasons.[75] These are the very sorts of aesthetic and educational practices woven into *A Girl of the Limberlost*. Stratton-Porter saw such artistic expression as a means of engaging deeply and creatively with nature. The highest calling of all, she believed, involved leaving a lasting legacy as an author, artist, or naturalist, who are "among the only ones who live beyond the grave in this world, the people who write books that help, make exquisite music, carve statues, paint pictures."[76] Echoing Comstock, she was convinced that the ethical and spiritual dimensions of appreciation for the natural world attain their finest expression through arts such as poetry and music, what Leopold would later term *cultural harvest*. Above all, the Nature Study Movement was predicated on the belief that unmediated interaction with nature contributed to ethical development and presumably a receptivity to the precepts of conservation.

Much like *A Girl of the Limberlost*, Stratton-Porter's next novel, *The Harvester*, addresses the ethical implications of development that result in the destruction of nature, as well as considering the virtues of stewardship—particularly the preservation of threatened native species. As the book opens, its protagonist, most often referred to simply as "the Harvester," is pondering whether to abandon life in the country "for the noise and grime of the city." But a lifetime on the land has instilled a powerful connection to place, a bond that provides a livelihood and one that ultimately defines him. While the overall romantic storyline is antiquated and even quaint by today's standards, it nevertheless delivers a serious message about environmental history and preservation to a remarkably large audience of avid fans.[77] Replying to contemporaries in the literary establishment who deemed such fiction implausible or saccharine, Stratton-Porter remained defiant, contending that her characters were based on people she knew, sometimes the composite of a few.

The Harvester resolutely resists allowing development of over six hundred acres he owns, though he knows the parcel could be profitably ditched, drained, and plowed. His dedication to preserving the many forest and wetlands acres is unwavering in the face of those neighbors who brand him a fool or fanatic for not cutting timber and cultivating corn. Beyond reserving wild ecosystems as natural habitat, the Harvester engages in stewardship, sowing wild rice in wetlands for birds to feed on among indigenous species such as blue flag, bulrushes, cattails, dock, and smartweed. Similarly, he encourages brambles of blackberry and raspberry to flourish for orioles and plants fruit trees such as cherry for robins. In contrast to the monotonous rows of monocultural cropland, he celebrates the native species thriving on his property with Edenesque extravagance, describing almost unimaginably verdant biodiversity:

> A marshy thicket of vines, shrubs, and bushes, the banks impassable with water growth. Everywhere flamed foxfire and cardinal flower, thousands of wild tiger lilies lifted gorgeous orange-red trumpets, beside pearl-white turtle head and moon daisies...while all the creek bank was a coral line with the first opening bloom of big purple mallows. Rank jewel flower poured gold from dainty cornucopias and lavender beard-tongue offered honey to a million bumbling bees; water smartweed spread a glowing pink background, and twining amber dodder topped the marsh in lacy mist with its delicate white bloom. Straight before them...the hill, crowned with gigantic forest trees.[78]

From his cabin he surveys a vibrant environment: "the fern and lily carpeted hillside, the majesty of big trees of a thousand years," an indigenous woodland community teeming with birds and wildlife, including killdeer and ducks, hawks and grebes, squirrels and weasels, as well as snakes in the thickets and bass in the lake.[79]

The Harvester also personifies aesthetics rooted in nature. He dabbles in folk arts, whittling candlesticks from native hardwoods, creating patterns based on natural forms encountered on his daily rounds, such as flowering vines, berries, and butterflies. With evangelical zeal he persuades his young bride to try painting from nature. Providing her with all necessary art supplies, he commissions her to paint the rare Yellow Emperor moth from life, as well as a luna moth as it emerges from its cocoon—just as Stratton-Porter was wont to do. He declares that working together they might "revolutionize decorative art...by introducing new, original designs" derived from organic forms in the natural world.[80] This desire to develop an artistic style predicated on nature might be likened to the nineteenth-century illustrations of natural historian, artist, and philosopher Ernst Haeckel (1834–1919), whose elaborate works—particularly the intricate depiction of marine species—inspired a movement in decoration and design that served as a precursor to Art Nouveau. He is also widely credited with coining the term *ecology*.

Like Stratton-Porter herself, the Harvester is a self-taught botanist who methodically studies books in a personal library brimming with works of natural history as well as herbal medicine.[81] Extensive use of curative herbs in the Americas is a longstanding tradition, of course, dating back to its indigenous peoples, whose ethno-botanical knowledge had accrued over millennia.[82] In fact, generations of frontier families in the Appalachians, such as the Cherokee, seeking a foothold in the cash economy of the whites, were able to supplement their income by selling wild ginseng root. During the nineteenth century, international trade in wild ginseng became highly lucrative, with a brisk market in Asia. Published accounts of fortunes to be made harvesting wild ginseng root in North America fueled rumors and a gathering frenzy of Gold Rush proportions. Such intensive harvest caused the extirpation of the plant in many regions from Georgia to Wisconsin where it had once thrived. To protect the resource, authorities in a number of states attempted to regulate harvesting along the lines of bag limits placed on hunting and fishing. Gradually, alternatives to gathering ginseng root have arisen, ranging from broadcasting seed in the wild for later harvest to outright domestication and cultivation of the crop. Yet wild roots have long carried a symbolic cachet of antiquity and authenticity and command a premium price to the present day.

The Harvester operates in this historical context, gathering medicinal plants in the wild to supply local customers such as physicians and druggists, as well as major pharmaceutical manufacturers nationally, while cultivating beds of particularly sought-after species on his own property. The tradition of gathering herbs had been passed down to him from his mother, initially to cover the cost of schooling and clothes. In order "to supply large concerns in the manufacture," the scale of his own operation had gradually grown as he processed an increasing variety of roots and seeds as well as leaves and bark."[83] The locals know him as the "Medicine Man," an eccentric and somewhat bookish recluse, prone to rattling off encyclopedic botanical knowledge. Plants of commercial interest include Canadian rock rose, catnip, foxglove, laurel, lilies, mullein, and willow. He is equally conversant in their individual uses, casually detailing in conversation how he "collect[s] catnip for infants and nervous people, hoarhound for colds and dyspepsia, boneset heads and flowers for the same purpose… [and] yarrow. I take the entire plant for a tonic and blessed thistle leaves and flowers for the same purpose."[84] Similarly, he can discourse on the optimal time for harvesting each one: "These I just have named I take at bloom time; next month purple thorn apple, jimson weed, and hemlock."[85] To gather successfully necessitates a keen appreciation for the progression of seasons, timing harvest precisely to optimize potency and maximize yield. Beyond raising, gathering, and processing medicinal plants, the Harvester experiments with combining varieties of herbal remedies "to allay the pain of humanity" and vows to devote himself to this noble calling.[86] To his mind, cultivating and preparing therapeutic herbs constitutes "the very greatest of all the great work a man can do."[87] "I don't see what bigger thing a man can do," he declares, "than to combine pure, clean, unadulterated roots and barks into medicines that cool fevers, stop chills, and purify bad blood."[88] When faced with his wife's dire illness, the only hope is to formulate remedies himself that druggists of the time could not provide.

Like many settlers eking out a subsistence existence on the frontier, the Harvester is situated at the cusp of two disparate cultural traditions: gathering and farming. While engaging in both of these activities, he also blends the two, such as when digging up wild roots to plant in his own garden, sowing seeds to propagate patches in neighboring woodlands, or systematically harvesting from mature plants that had produced seeds, so he can replant while gathering roots. His pride and joy, "acres and acres of wild ginseng, seven years of age and ready to harvest," represents in his eyes a treasure trove.[89] Cultivating native medicinal plants so extensively on his own land, he realizes, serves as a hedge against their eradication in the surrounding region due to development, particularly monocultural crop production. "The place is an experiment. When medicinal herbs, roots, and barks became so scarce that some of the most important were almost extinct, it occurred to me that it would be a good idea to stop travelling miles and poaching the woods of other people, and turn [my] land into an herb garden."[90] Accordingly, as a defense against extirpation, "he went on steadily enlarging his species, transplanting trees, shrubs, vines, and medicinal herbs from such locations as he found them to similar conditions on his land. . . . cultivating these beds and hunting through the woods on the river banks, government land, the great Limberlost Swamp, and neglected corners of earth for barks and roots," vowing to "remain at home to harvest golden seal, mullein, and ginseng."[91] Keenly aware of local environmental history, he recognizes how fertile lowlands were the first to be converted to cropland, causing decline in the biodiversity of indigenous species.

The native species the Harvester seeks to reintroduce and cultivate are those extirpated to the point of extinction locally—a direct parallel with Stratton-Porter's own extensive transplanting of wild specimens at her second homestead at Limberlost North in northern Indiana. Much as the author herself had done, the Harvester establishes patches of plants on his own land that he christens Medicine Woods. Similarly, the Harvester's efforts to propagate particular plants to enhance wildlife habitat—such as alder trees, grapevines, raspberry patches, and wild cherries for the benefit of birds—closely resemble the author's own endeavors. Stratton-Porter sowed native species of plants that could attract and sustain populations of migratory birds as well as the butterflies and moths she so avidly studied, and in the process preserved a microcosm of the region's historic biodiversity.

Hoping to inspire readers to pursue similar stewardship themselves, Stratton-Porter advocated that homeowners everywhere landscape by planting evergreens such as cedar, cypress, and juniper for nesting habitat, as well as fruit trees for forage. She reintroduced indigenous species adapted to regional climate and soils long before such initiatives became more widespread. In order to attract magnificent moths, she advised readers in 1912, it was simply necessary to provide the right kinds of plants as habitat: "trees, bushes, and vines liked by caterpillars are apple, quince, plum, and pear of the orchard; gooseberry and currant of the garden; lilac rose and snowball of ornamental shrubbery; and among the wild trees, oak, hickory, walnut, cherry, poplar, maple, tulip, elm, willow, sassafras, and alder."[92] She also advocated restoring absent populations by introducing the cocoons of desired species of moths then widely available for sale. The ecological memory operating in such a natural sanctuary parallels the cultural memory preserved in her books. As ecocritic Sydney Plum contends, "fence lines are the repository and haven of survivors of a too greatly harvested abundance.... Similarly Stratton-Porter's words and photographs are like the fences: a line of defense against irreparable loss."[93] After a lifetime of exalting natural beauty in the field, Stratton-Porter had been moved to engage in hands-on stewardship. She championed the cause of environmental restoration as a hedge against seemingly relentless destruction of the natural world: "We must make reparation or we must meet disaster.... We must replace, as lies in our power, our timber lands for the sake of the beauty which is absolutely necessary to the development of the soul to the fullest enjoyment of life, we must protect the natural beauty remaining to us, the flowers, the birds, the fish."[94] Such ideals were not merely a philosophical stance for her, but a vision that she enacted with inexhaustible vigor. In an unprecedented gesture of restoration, she personally transplanted native species—estimated at over seventeen thousand plants in all—to semi-feral gardens on her property that she christened Wildflower Woods.[95] In the process, she hoped to save many of her favorite wildflowers—such as blue bellflower, cardinal flower, foxfire, and mandrake—that the expansion of cropland had diminished to near extinction in the region. In short, "the urgency to preserve native species was uppermost in her mind."[96] She conceived of the project as a living museum that she hoped to bequeath to the state of Indiana one day, preserving a glimpse of pre-settlement ecosystems for future generations. Her wish would be fulfilled decades after her tragic passing in 1924, at the height of her powers as a writer, photographer, and filmmaker.

Stratton-Porter pursued this initiative to the very end of her life, hoping that the state of Indiana would take responsibility for managing the entire property as a botanical preserve in perpetuity. She wrote to state officials, describing the extensive plantings at Wildflower Woods and requesting that they be "taken over by the state for a permanent wild flower garden and bird preserve."[97] The letters she wrote advocating the creation of a public park reveal how the purpose for this project was driven by her overall vision of conservation and her keen desire to preserve indigenous species: "Every bird native to the Central States homes there. I personally have released hundreds of moths and butterflies there, and my records show that I have planted very close to twenty thousand trees, shrubs, vines and bushes, besides sowing pecks of wildflower seed in order that I might save here for the nature lovers of the state, specimens of the rarest and most beautiful wildflowers."[98] Such an ambitious plan—let alone personally seeing it through to fruition—is a largely unsung component of her overall environmental legacy, one that anticipated Leopold's celebrated restoration of ravaged farmland on the Sauk River in southern Wisconsin.

We cannot be certain how Stratton-Porter would sort out the conundrums surrounding *rewilding* today, though she surely would have championed the principle of engaging private landholders and homeowners in the process. How would she view commercial propagation of "nativars" such as coneflowers, for example, purported to support populations of birds and butterflies? Intimate acquaintance with local populations of native wildflowers convinced her of their vulnerability in the face of landscape change, as well as their adaptability. Given her serious study of biology and natural history, moreover, there is reason to believe that she would have recognized the risks of subverting processes of natural selection through approaches to cultivation based on cloning or genetic modification, and intuited how reducing genetic diversity within a species might jeopardize the resilience of self-propagating populations.

Stratton-Porter's pleas for preservation of Limberlost Swamp outside Geneva, Indiana, have been heeded belatedly by those endeavoring to reclaim significant swaths of the original swamp today. Moreover, her vision of restoration continues to inspire contemporary literary naturalists. As Scott Russell Sanders describes, "She lured me to this place by her photographs and words to Loblolly Marsh to see a remnant of the vast, magnificent, vanished wetland."[99] Meadows have been replanted with native species of prairie grasses. Cattail marshes are teeming once again with flocks of red-winged blackbirds. Indigenous species of trees rebound in boggy patches: cottonwood and maple, sweet gum and sycamores, along with willow and several varieties of oak. "This healing fragment of original swamp," Sanders reports, now offers habitat for birds, including barn swallows, bluebirds, cardinals, chickadees, crows, ducks, goldfinches, killdeer, phoebes, sparrows, turkey vultures, and wrens; while beaver, deer, fox, possum, and raccoon return to forage and prowl, a striking instance of ecological memory in the biological sense. Wild communities of indigenous species have been reestablished with rapidly increasing biodiversity as natural ecological processes again take hold.[100]

NOTES

1. Plum, *Coming Through the Swamp*, ix.
2. Armitage, *Nature Study Movement*, 210.
3. Stratton-Porter also expressed great admiration for the writings and exploits of John Muir in a letter (March 6, 1910) to Enos Mills, a naturalist inspired by an encounter with Muir and an early advocate of Rocky Mountain National Park. In fact, Stratton-Porter's personal library contained at least two of Muir's books, *My First Summer in the Sierra* and *The Story of My Boyhood and Youth*, as well as Mills's *Waiting in the Wilderness* and Samuel Young's *Alaska Days with John Muir*.
4. Thoreau, *Walden: A Fully Annotated Edition*, 306.
5. The Nature Study Movement, to be detailed later, was a widespread educational initiative in American schools that gained influence throughout the country during the first decades of the twentieth century. It emphasized direct experience of the natural world.
6. Stratton-Porter, "All Together, Heave!," 149.
7. Stratton-Porter, *Tales You Won't Believe*, 163.
8. Richards, *Gene Stratton Porter*, 19.
9. Stratton-Porter, *Homing with the Birds*, 8.
10. Leach, *Butterfly People*, xvii.
11. Obuchowski, *Field O' My Dreams*, 7.
12. Stratton-Porter, *At the Foot of the Rainbow*, 13.
13. Ibid., 33.
14. To put this achievement and her popularity into perspective, consider that of fifty-five books that sold more than a million copies in the fifty years between 1895 and 1945, she single-handedly authored nearly 10 percent of them (Armitage, *Nature Study Movement*, 246).

15. Both these books were turned into movies; in fact, there are five film versions of *A Girl of the Limberlost.*

16. Birkelo, "Gene Stratton-Porter," 26.

17. Armitage, *Nature Study Movement*, 168.

18. Birkelo, "Gene Stratton-Porter," 7.

19. Obuchowski, *Field O' My Dreams*, 12.

20. Stratton-Porter, "My Work and My Critics," 152.

21. Obuchowski, *Field O' My Dreams*, 3.

22. Stratton-Porter had copies of a number of Thoreau's books in her personal library, including *A Week on the Concord and Merrimack Rivers* (1849), *Walden* (1854), *Excursions* (1863), *The Maine Woods* (1864), *Cape Cod* (1865), *Winter* (1888), *Autumn* (1892), and *Miscellanies* (1894).

23. Stratton-Porter, *Moths of the Limberlost*, 3.

24. Stratton-Porter, *At the Foot of the Rainbow*, 30.

25. Stratton-Porter, *Music of the Wild*, 250.

26. Although ongoing destruction of the wetlands troubled her deeply, Stratton-Porter took advantage of the proliferation of new logging roads to make frequent collecting and photographic trips, gaining access to progressively remote parts of the swamp.

27. Plum, *Coming Through the Swamp*, xxvi.

28. Stratton-Porter, *Music of the Wild*, 127.

29. Stratton-Porter, *Homing with the Birds*, 66.

30. Stratton-Porter, "Hidden Treasures," 32.

31. Stratton-Porter, *Music of the Wild*, 335.

32. Among her contemporaries, only a handful of bold field biologists speculated publicly on the cumulative ecological impact of wildlife population crashes, such as William Temple Hornaday, whose seminal tract *Our Vanishing Wildlife* (1913) railed against the specter of extinctions and urged conservation policies that could prevent them.

33. Leach, *Butterfly People*, xviii.

34. Stratton-Porter, *Homing with the Birds*, 365. In the century since, proprietary chemical compounds coupled with the genetic engineering of seeds have led to increasing dependence on petrochemical herbicides and pesticides. At the very same time, natural insect control by songbirds has been jeopardized by population crashes due in large part to pervasive habitat fragmentation.

35. Stratton-Porter went so far as to speculate that following the discovery of solar systems orbiting stars other than the sun, it would seem reasonable to assume that life would evolve on some of those planets and in much the way it has on Earth. Writing in 1923, she even hypothesized that since they had "formed practically of the same substances as our world; they must evolve life in the same way," the very sort of conjecture astronomers continue to voice today (Obuchowski, *Field O' My Dreams*, 138).

36. Stratton-Porter, *Birds of the Bible*, 19.

37. Ibid., 24

38. Stratton-Porter's personal library included at least three volumes of Darwin's works pertaining to botany, including *Different Forms of Flowers and Plants*, *Power of Movement in Plants*, and *Insectivorous Plants*.

39. Stratton-Porter, *Homing with the Birds*, 321.

40. Obuchowski, *Field O' My Dreams*, 7.

41. Stratton-Porter, *At the Foot of the Rainbow*, 47.

42. Meehan, *Lady of the Limberlost*, 228.

43. Stratton-Porter, "A Little Story of Her Life and Work," in *At the Foot of the Rainbow*, 47.

44. Stratton-Porter, unpublished archived letter, May 4, 1922; Stratton-Porter archival material is housed at the Gene Stratton-Porter Historic Site in Rome City, Indiana.

45. An exquisite portrait of her daughter Jeanette that Stratton-Porter rendered in pencil, for instance, is displayed to this day in the parlor of the family's Limberlost home site at Geneva, Indiana.

46. Plum, *Coming Through the Swamp*, xvi.

47. Stratton-Porter, *Homing with the Birds*, 346–47.

48. Stratton-Porter's reception by her contemporaries is reminiscent of the disparaging treatment Audubon received, being discredited as amateur by the reigning academics of his day. Gender was also a significant factor in her professional reception in an age of "polite botanists," when women were still being treated dismissively in the sciences.

49. Plum, *Coming Through the Swamp*, 91.

50. Meehan, *Lady of the Limberlost*, 143.

51. Stratton-Porter, "Hidden Treasures," 60.

52. Ibid., 36.

53. Stratton-Porter, *Homing with the Birds*, 67, 310.

54. Armitage, *Nature Study Movement*, 153.

55. Stratton-Porter, *Moths of the Limberlost*, 111.

56. Similarly, the protagonist in her novel *The Harvester* was reportedly a composite figure inspired by Thoreau as well as the author's own father.

57. Stratton-Porter, *Girl of the Limberlost*, 82.

58. Ibid., 168.

59. Ibid., 273.

60. Ibid., 167–68.

61. Ibid., 55.

62. Ibid., 237–38.

63. Ibid., 61.

64. Ibid., 98.

65. This focus on grounding the life sciences in intimate personal acquaintance with local places corresponds to what environmental historian Robert Kohler refers to as "residential research."

66. Ibid., 242.

67. While it is tempting to speculate that the character Elnora Comstock was named as a tribute to the Nature Study Movement's leader, there is no definitive evidence of any relationship between Anna Botsford Comstock and Stratton-Porter, such as letters between the two.

68. Armitage, *Nature Study Movement*, 206.

69. Anna Botsford Comstock was the first female professor in the history of Cornell University. Comstock's influential *Handbook of Nature Study*, published in 1911, a work that emphasized natural history in school curricula at every level, provides a historical context for the pedagogy advocated by the Nature Study Movement—much like that portrayed in the novel.

70. Stratton-Porter, "Hidden Treasures," 62.
71. Ibid., 33.
72. Armitage, *Nature Study Movement*, 206.
73. Ibid., 13.
74. Ibid., 4.
75. Ibid., 157.
76. Stratton-Porter, *Girl of the Limberlost*, 49.
77. *The Harvester* struck a particularly deep chord with readers. Stratton-Porter eventually received countless letters from young men inspired to emulate the book's protagonist by finding reputable ways of earning a decent, morally upright living while residing in the countryside. In fact, though intended as an inspirational model of traditional personal virtues and civic values associated with agrarian living, the work circulated widely in prisons.
78. Stratton-Porter, *Harvester*, 125.
79. Ibid., 153.
80. Ibid., 91.
81. Stratton-Porter's own library included a generous assortment of scientific texts of the period, addressing botany (particularly native species of trees and wildflowers), entomology (particularly moths and butterflies), ornithology, astronomy, and geology.
82. Russell, *Indian New England*.
83. Stratton-Porter, *Harvester*, 15.
84. Ibid., 93.
85. Ibid.
86. Ibid., 166.
87. Ibid., 181.
88. Ibid., 23.
89. Ibid., 128.
90. Ibid., 135.
91. Ibid., 16.
92. Stratton-Porter, "Hidden Treasures," 63.
93. Plum, *Coming Through the Swamp*, xxiv.
94. Stratton-Porter, "All Together, Heave!," 173.
95. Wallace, "Afield with Gene Stratton-Porter," 109.
96. Richards, *Gene Stratton Porter*, 36.
97. Stratton-Porter sent photographs of the Limberlost North estate with a letter dated September 13, 1923, to the commission charged with considering transferring her holdings at Sylvan Lake to the state—namely, Wildflower Woods plus the Limberlost North home site on 120 acres—and soliciting the assistance of Commissioner Hohenberger as well as Col. Richard Lieber, who would become instrumental in establishing Indiana's system of state parks, which did not actually acquire the Stratton-Porter property until decades later in 1946.
98. Stratton-Porter, unpublished archived letter, April 22, 1924.
99. Sanders, *Conservationist Manifesto*, 145.
100. Ibid., 145.

3

To Live in the Wilderness as a Wild Creature Myself

PAUL ERRINGTON (1902–1962)

This is the testimony of the marsh:
Life directs all its power to one end and that is to continue to be.

—KATHLEEN DEAN MOORE, *Holdfast*

Accounts of landscape change in the Upper Midwest by pioneering ecologist Paul Errington can be read as capsule environmental histories, yet also represent a cautionary tale: how unchecked development had rapidly stripped so many verdant wetlands of their especially rich biodiversity. Along with other wildlife biologists in the vanguard of ecological thought at its inception during the first half of the twentieth century, he helped to fashion a vision of conservation, particularly in response to the ongoing transformation of prairies for agricultural use.[1] What seems prescient is his recognition that he had been privileged to glimpse as a young man a level of abundance as fleeting as it was extraordinary, and as awe inspiring to him as the natural bounty of North America had been to the first generations of Europeans to explore and colonize the continent. He grasped that he was witnessing an epic and possibly irrevocable loss of wildlife throughout much of the region spanning grasslands across the Great Plains. Such destruction exacted a monumental cost for humanity as well, since the solitude, solace, and meaning natural landscapes can provide us would be lost to posterity. "Wilderness and what it stands for," he declared, "could mean still more in terms of human tranquility and reflection in an era of artificiality and unrest," an increasingly poignant admonition since it was written in the 1950s.[2]

Of Men and Marshes, Errington's best-known and most compelling work for general readers, is rooted in an ecologist's deep regard for the workings of nature. Not only had he observed abundance preceding precipitous decline, he appreciated the magnitude of what was at stake environmentally in the long run, making *Marshes* "at once history and prophesy."[3] Based on his extensive experience as a field biologist studying the dynamics of the region's wildlife populations, he advocated preserving the integrity of verdant wetland ecosystems. The volume represented at once a culmination of his career as a scientist and an important departure in extending his reflections as a wildlife biologist by staunchly advocating conservation. "What has counted most to me in wilderness values," he declared in the posthumously published *Of Wilderness and Wolves*, "has been the living and non-living constituents of a given tract of wilderness."[4] While he was writing generations before the idea of *wilderness* became contested, his work clearly anticipated many of the issues still debated today. In later years especially, he championed preservation of wildlife and wetland habitat with an ever greater sense of urgency, understanding that the survival of certain species hung in the balance.

By the end of his career, Errington had visited, observed, and studied wetlands throughout North America, including in states as far flung as Florida, Georgia, Idaho, Maryland, Montana, Nebraska, North Dakota, South Dakota, and Utah, as well as several provinces in Canada. Even places as austere as the Dakota Badlands had moved him to "a reverence for the antiquity and integrity of life."[5] As a Guggenheim Fellow in the late 1950s, he became acquainted with glaciated landscapes of Scandinavian nations, including Denmark, Finland, Norway, and above all Sweden. Still, his overall understanding of ecological processes and theories about interdependence—especially relationships between predator and prey—were grounded in a lifelong exposure to landscape change, particularly in Minnesota and the Dakotas. Accordingly, his work as a field biologist was primarily devoted to mammals that inhabit marshes of the Upper Midwest. His six books provide intimate depictions of the behavior of a wide variety of wetland species, which ultimately led him to question (as Aldo Leopold did) the conventional wisdom about the ecological role of predation and the efficacy of predator control. Federal and state policies promoting eradication, he realized, were predicated on flawed assumptions about the effect of predation on wildlife populations and typically overlooked the ecological impact of landscape change on habitat. At a time when the extermination of major predators such as wolves was standard operating procedure among wildlife managers, he sought to elucidate critical interrelationships between members of wetland communities as well as complex webs of interdependence among species. Moreover, he understood that predator and prey that share habitat may in fact have *coevolved*, adapting together to the dynamics of a particular environmental niche.

Errington expressed tremendous regret that by the 1950s, wolves had long been "extirpated over such a tremendous portion—nearly all—of the range that they once held in those parts of the United States lying south of the Canadian border," and for that reason he argued passionately for "protect[ing] the remnant populations that persist in wilderness tracts," not least of all because "of how superbly they represent wilderness."[6] Although vanquished from much of their historic range across North America, in a few places isolated packs survived, continuing to prey on the same native species that had sustained them for millennia, in a relationship shaped by monumental landscape change: "While the continental glaciers advanced and receded," he observed, "modern wolves and their wild prey adjusted to each other and underwent their own respective evolutions."[7] Viewed from the perspective of deep time, he realized, the landscape is ever changing. In *Marshes*, he invoked the history of glaciation that had shaped the landscape, carving out pothole lakes and wetlands. "The duration of a marsh, individually, [is] but a geological moment," he wrote, "with moments recurring over and over again to produce geologically ephemeral new marshes."[8] Moreover, he grasped that the abundance he had witnessed in his youth—the consequence of an ancient conjunction of geological and ecological processes—was becoming increasingly rare and precarious.

By the middle of the twentieth century, Errington already saw evidence that some species of migratory waterfowl such as redhead and canvasback ducks, which had been declining for decades, were threatened in parts of their historic range throughout the Upper Midwest. These wild populations, he believed, "truly belong on our marshes... species that have been there since before the white man came," as opposed to the domesticated ducks increasingly being raised for market.[9] "As a conservationist, I have been disappointed," he lamented, "in the losses of outdoor values that have occurred since my youth... by the wastage of woodlands, wetlands, and grasslands. I may wonder if mankind's overrunning and despoilation of the earth will stop short of ruin."[10] Consequently, he felt it imperative to preserve as many different types of habitats as possible, precisely because myriad species have adapted over eons as a part of every ecosystem, each with its own "natural biological equations."[11] For example, preserving extensive wetlands was essential to sustaining robust populations of waterfowl across the continent. Moreover, since the nesting habitat preferred by each species varied widely, supporting biodiversity necessitates protecting many kinds of wetland ecosystems. Indeed, doing so seemed an ethical obligation to him.

Errington described marshes and lakes of South Dakota utterly teeming with waterfowl during his youth, when the region still supported populations of historic proportions: "in variety and numbers possibly not found on Earth since the decade following the First World War."[12] "The mallards were always in the background," he wrote, "in their incredible millions of the 1920s in the South Dakota wetlands."[13] While he marveled at the "spectacular abundance of many species...up to three-quarters of a million red-winged blackbirds...in a single part of a large cattail marsh,"[14] he realized that even such a great multitude of birds "hardly compared" with the numbers his grandparents would have witnessed when they homesteaded in the Tetonkaha Hills, presumably named for the great herds of buffalo that once wandered there. Still, he reveled in the verdant wildlife of the region during his own youth, describing the immense flocks of ducks he witnessed at Lakes Albert and Tetonkaha as "incredible as measured by present-day experience."[15] In fact, he recollected many years later, when he first camped there at the age of eleven, Lake Albert was a place "then known for its wonderful abundance of game."[16] While he acknowledged that previous generations of settlers had grazed livestock and even cleared a few fields, the area as a whole remained in his estimation largely "wild." Scattered across the backcountry tracts of South Dakota were the "weathering ruins of buildings and farming machinery and windmills left by the too-many people who formerly had tried to live on land that should not have been treated as farmland."[17] Unlike Errington's more technical publications, *Marshes* makes its "sole topic the beauty and biological richness of the midwestern glacial marshlands" and, as ecocritic Matthew Sivils observes, "eloquently expressed the dire need to save the endangered wetlands from drainage and other forms of human encroachment."[18] In the epilogue of that volume, he lays out personal principles of environmental ethics. Like literary naturalists before him and since, from George Perkins Marsh in the nineteenth century, to his contemporary Aldo Leopold, and to Scott Russell Sanders in our time, he viewed the preservation of wildness as a measure of civilization itself. "A civilized attitude would be to try to preserve a good deal of Nature in as natural condition as we can," he contended, "if only for the sake our own mental health."[19] Leopold had expressed similar sentiments, claiming that civilization is better measured by how well the ecological integrity of the environment has been preserved than by the extent of metropolitan development alone; indeed, in their estimation, the highest expression of "culture" is to be found in an ethical commitment to habitat preservation.[20]

Errington devoted the final decade of his life to writing one last collection of striking essays: *Of Wilderness and Wolves*, completed in 1961, though not published until 2015. While inspired in part by Aldo Leopold's *A Sand County Almanac*, *Of Wilderness and Wolves* added an enduring dimension to his legacy as a literary naturalist. There were many parallels between the lives and careers of these two midwestern ecologists and literary naturalists.[21] They each grew up as avid outdoorsmen, hunters, and students of wildlife. Their views on conservation ultimately matured: from those of wildlife managers charged with protecting populations of game animals favored by hunters to a more inclusive vision that raised compelling questions about the significance of wetlands ecologically and culturally. "Like Aldo Leopold, Paul Errington ventured into the realm of philosophical advocacy for natural resources and wildlife... questioning basic societal norms... [and] illuminating the life and essential values of wetland environments."[22] Indeed, the two shared a commitment to preserving the integrity of natural ecosystems for posterity: as a source of inspiration as well as for scientific inquiry. In the end, both became advocates for conservation, lamenting the reduction of habitat and declining wildlife populations and protesting the destruction of thriving natural ecosystems for the sake of development. The aim of *Of Wilderness and Wolves*, he confided, was "to regard wolves for what they are without mythology and sensationalism... distinguishing between wolves of fiction and wolves of reality."[23] In truth, these late essays seek to achieve a far more profound outcome: to "rethink, revalue, and preserve our increasingly threatened natural world."[24]

What distinguished Errington, as well as Leopold, from many other scientists of their generation was having spent such extensive time observing in the field long before the start of their careers. As a young man, Errington had hunted and trapped in the wetlands of east-central South Dakota adjoining Lake Tetonkaha. In fact, as a fledgling academic in the 1930s, as Sivils explains, "he quickly became a major force in the development of the emerging discipline of wildlife ecology, spending decades conducting research upon the very marshland ecosystems he loved as a young trapper... particularly interested in the intricacies of predation as it played out upon the marshland stage."[25] He described his family's farm as being "right in the midst of some first-of-the-century wolf country.... There were buffalo wolves—the real lobos—and much wild land."[26] Indeed, in those days the howling of wolves still "remained a possibility on a South Dakota night," though he conceded that none had been seen in the immediate vicinity of the homestead recently. He acknowledged how those landscapes had made an indelible impression on his imagination: "the wilderness, and the wilder places of backwoods, prairie marshes, and prairie streams, and the rugged country and great expanses of the high plains."[27] Formative experiences in such undefiled places shaped his lifelong convictions about preserving the natural world. In fact, beginning in South Dakota, his "firsthand knowledge of the lifecycles and habits of marshland creatures, of the interrelationships between these organisms and their habitat, set him apart...time spent in the marshes proved crucial to the development of his ecological consciousness, to his appreciation of the complex interrelationships found in the marshland environments."[28] His ecological perspective had in fact developed throughout his lifetime; he recounted how "early outdoor life instilled [him] with appreciation for the integrity of nature...associations of plants and animals and physical environments that were time tested."[29]

Modern readers may have trouble reconciling Errington's professed affection for wildlife with his passion for shooting. His path to becoming an ardent environmentalist, much like that of his contemporary Aldo Leopold, or even John James Audubon before them, had begun with gun in hand. Indeed, Audubon's credibility as a conservationist has been contested given his infamous penchant for collecting what seems, by today's standards, an excessive number of specimens. In a throwback to such nineteenth-century ornithological practices, Errington was briefly licensed to shoot protected species of birds as specimens for a museum collection at South Dakota State University. By the middle of the twentieth century, he had become aware that others might see a contradiction where he himself did not—between a love of hunting and the ecological and ethical imperatives of preserving wildlife. In fact, his memoir ends on a poignant note as he anticipates this very issue. He offers not so much an apology as an explanation, though tinged with regret: "Some people may wonder how I may claim that my shooting of all those ducks was an expression of love for the ducks and for what they and their wildness and freedom stood for in my mind. That was what it was. That hunt was a form of saying goodbye."[30] This experience, he adds, was "ritualistic." We might speculate that these words of leave-taking represent more than a young man's nostalgia for carefree adventures in the field at the very moment he decided to pursue a career as scientist and academic. The elegiac tone also signals his grasp of environmental history and contains an implicit lament for the natural abundance of North America that he understood had already been jeopardized.

Errington was born into the vestiges of frontier hunting culture in the Upper Midwest at the dawn of the twentieth century, and his career as a wildlife biologist also straddled eras in the natural sciences: "Errington can be seen as a scientist who embodies both natural history traditions as well as 'modern' propensity to develop and emphasize theory."[31] Being at the cusp of profound historical change culturally and professionally, as well as environmentally, gave him the intellectual leverage of a diachronic perspective. Interpreting landscape change, he realized, inevitably entails triangulation across time, as well as an interdisciplinary vision of how various types of ecological alteration might impact the population dynamics of wildlife. This was the sensibility he brought to reading environmental history.

In *The Red Gods Call,* Errington describes how experiences afield while still in high school had given him his first inklings of ecological principles. While "such stories are a literary trope: a foretelling of the future scientist in the curious youth," Robert Kohler notes, "they also bear witness to experience of actual places and of a particular time in the environmental history of North America. They are stylized recollections of what it felt like to inhabit the twilight zone, where wild (or wildish) nature was experienced."[32] Toward the end of his life, Errington recounted that as a boy he had already glimpsed the "rules of order behind natural relationships," recognizing intuitively the complex Darwinian processes of adaptation to habitat.[33] That early passion as an outdoorsman was reflected later in an affinity for fieldwork as a professional wildlife biologist of the highest order.

Errington recollects in his memoir how he had reveled in ecological vitality of the Upper Midwest during his boyhood, as "our family farm at Lake Tetonkaha could be, on occasion, just about full of all kinds of wetland wildlife native to east central or northeastern South Dakota."[34] He characterizes the landscape as "virgin prairies" and describes the migration of waterfowl in "uncountable thousands." Moments are recounted with a visceral immediacy: "I became aware of ducks by the thousands and thousands, flying all over and around me."[35] The Paul L. Errington Memorial Marsh near his grandparents' homestead on Lake Tetonkaha in South

Dakota commemorates a tract where he first developed a personal sense of ecological reckoning while hunting and trapping in his youth.

Errington's memoir also recounts formative outdoor adventures in progressively remote territory in Canada, then still sufficiently undeveloped to support thriving populations of furbearers. From a tender age, he romanticized the era of mountain men trapping in rugged terrain beyond the reach of frontier settlement. He was infatuated with the tales of "old-timers... of prairie fires and storms and lobo wolves and Indians, of the breaking of prairies and grasshopper plagues."[36] For a time, he even entertained the ambition of making a living at writing fiction centered on such fanciful exploits in the Far North, along the lines of Jack London's classic Alaska adventure stories *White Fang* and *Call of the Wild.* In fact, he described *Call* as "the most fascinating and disturbing book" he had read as a child. "Emotionally," he declared, "it turned me inside out" and fueled fantasies of pursuing treacherous backcountry enterprises of his own in isolated and frozen places.[37] Yet this calling to become an author of outdoor adventure stories proved to be short-lived, no more tenable than the outmoded venture of trapping itself. Still, the impulse reflected a deep desire for the primal experience of engaging with animals in the wild, and to participate directly by hunting and trapping. In fact, he declares his aim at that time to have been "to live in the wilderness as a wild creature myself as a predator along with the mink, the lynx, the bear, and the wolf."[38] While dangers such as lethal cold initially only heightened his sense of adventure, such "romantic fancies of the heroic trapper-naturalist" were shattered by a "brush with death in a real-live wilderness" the winter he spent trapping in Canada.[39] Even his later popular science writing reveal just how much he cherished those early experiences. Writing decades afterward, he wistfully described "memories of long winters, of northern lights, and sinking frost lines, of drifting snow over plains and ice, of stillness of spruce and cedar swamps, of snare wire and Newhouse traps, of weariness, loneliness, beauty."[40] While earnings as a fledgling trapper had often been meager, Errington treasured recollections from such far-flung and rugged places: these were just the kind of unsettled landscapes that established in his mind a baseline for natural abundance and ecological integrity. Indeed, his deep appreciation for environmental history stemming from those early experiences hunting and trapping inspired and informed his career as a field biologist.

By the first half of the twentieth century, much of the Midwest was dominated by a matrix of working, agricultural tracts, as Kohler explains:

"Prairie marshes were islands of 'wilderness' in a settled landscape. It was a fleeting moment in the region's environmental history between lightly settled 'inner frontiers' and an industrialized cornbelt landscape, and Errington's generation was about the last to experience it."[41] While quick to acknowledge the value of agriculture, he protested the ongoing loss of wetlands and called for preservation of those that remained, holding out hope that one day still others might be restored. He argued, furthermore, that preserving even fertile farmland and mature forests in their natural state should sometimes supersede development. He called for "the preservation of some places with the best timber or soil...that could easily be plowed, and kept plowed, some marshes that could be turned into cornfields."[42]

Throughout his life Errington witnessed the ongoing and often abrupt transformation of the Dakotas and Minnesota, as advances in technology allowed a greater range of wetlands to be drained for conversion to agricultural use as cropland. Farther east in Indiana, such development had commenced generations earlier. While five and a half million acres of wetlands had initially presented a formidable impediment to settlement, by 1882 only one million acres remained statewide after the installation of tens of thousands of miles of drainage tile.[43] The million-acre Kankakee Marsh, for instance, was methodically drained during the late nineteenth century, shrinking an exceptionally verdant wetlands complex to just 5 percent of its original extent, and reducing migratory bird populations *for the whole of North America* to an alarming degree, by some estimates to one-fifth their former size. Such "improvements," cloaked in the rhetoric of progress, came at a monumental environmental cost.

Errington recognized that wetlands provide especially vital habitat for a host of wildlife: in extreme northerly climes, for instance, he observes that "the marsh-edge visitors include wolves, bears, otters, skunks, and minks; in the western high plains, coyotes, badgers, skunks, and minks; in the north-central states, foxes, weasels, raccoons, skunks, and minks."[44] Indeed, he viewed species biodiversity in each region as an indicator of the environmental integrity of its wetland ecosystems as a whole. In his estimation, during the first quarter of the twentieth century there were still tracts of "wilderness" in the Upper Midwest, such as portions of the Big Bog in northern Minnesota, places sufficiently "remote from both land-clearing settlers and lumbering operations."[45] He recounts a litany of species encountered in its deeper reaches, including beavers, coyote, deer, fishers, grouse, lynx, mink, moose, owls, skunks, squirrels, and wolves. He was convinced that such "wildlife brought both intrinsic beauty and intrinsic value to the ecosystem."[46] Certain species, he realized, including major predators such as wolves, lynx, and owls, have long had symbolic resonance by being associated with wilderness habitat. In terms of wildlife management, he understood that the survival of these iconic species often depended on protecting the ecological integrity of large tracts of territory, even areas previously developed, which ideally included varied habitat enhanced by edges and ecotones.

Errington was under no illusion that the Big Bog in northern Minnesota, albeit forbidding, was "pristine" or even free from signs of previous human use. When traveling in the backcountry he often noted with relish that moment of crossing beyond the last decaying trace of previous human habitation. Returning to the region forty years after having trapped there as a young man, he recounted its environmental history of exploitation as well as subsequent efforts at preservation. In the process, he contemplated the fate of such recovering habitat in a way that resonates with present-day debates about the meaning of "wilderness" in the wake of landscape change: "Long ago the lumberman took the big pines, and old tote roads and other reminders of human activities are still visible even in some of the remote parts of the park.... But the park does include and is surrounded by great expanses of wilderness—wilderness properly designated as such even though it cannot rate as primeval wilderness. It has the creatures native to northern wildernesses, including wolves and wolf prey."[47] Moreover, he was heartened to recognize how the forces of ecological memory in the biological sense had begun to reassert themselves, through forest succession and the return of historically indigenous species of plants and animals. "The Big Bog was no longer subject to the exploitation and promotional schemes of earlier in the century.... After its withdrawal from agricultural use...it had regained some of the wilderness features lost during the boom period of human settlement. Perhaps," he concluded, "it might be considered, battered as it was, a triumphant wilderness."[48] He celebrated its increasing biodiversity: a regional suite of "age-tested native" species, including deer, foxes, grouse, hares, skunks, and weasels, plus the presence of moose—but above all wolves—"reassurance that the Big Bog was still wilderness, despite the decades of human use.... Wolf country this had been since before the white man came, and wolf country it still was."[49]

Yet even in the most secluded settings, he realized, ancient ecosystems were eventually susceptible to the pressures of human intervention and development. In *Of Wilderness and Wolves*, Errington enumerated an array of ecological factors implicated in dramatic changes in the range of many species, even in seemingly isolated locations. Spikes in wildlife populations, he realized, are inevitably followed by corresponding cycles of decline—a ledger of checks and balances that has been fashioned gradually, in fits and starts, since the distant past: "Whether the elimination of the over-produced occurs through predation or not, a natural shaking-down to fit wild populations into their habitat must have the sanction of evolution."[50] Local extinctions, he recognized, were typically caused by a combination of several factors, what he termed "the culminating symptom of biological impossibility.... The edge of a species range had ceased to be range at all."[51] For example, the "last free living remnant" of the woodland caribou in the United States, once indigenous to the Big Bog region of Minnesota, had disappeared entirely. However, declines in game species prized by hunters, he reasoned, cannot be attributed to predation alone since population dynamics are equally subject to a host of other influences: "Changes in human land use, climatic cycles, and dominance of plant successions," he pointed out, "may push one species up and another down."[52] Much as the heath hen had succumbed to excessive market hunting, for example, prairie grouse might become vulnerable to extinction, especially if faced with a number of threats at once. Wildlife ranges, he concluded, were ultimately ephemeral: intrinsically dynamic and forever in flux. Accordingly, reading the ecological configuration of a landscape at any given point in time entailed reflecting on its environmental history.

Errington was fascinated with places where the environmental history of human use was reversed through recolonization by species once native to the region. He described land that had been settled, farmed, and later abandoned beginning to revert to a wilder state, including stretches of Big Bog that "did not much resemble the wilderness the white man had found there, but [had] regained some of the wilderness features lost during the boom period of human settlement."[53] Revisiting the Tamarack River in Minnesota's North Woods where he had trapped twenty-eight years earlier, he recognized fundamental changes precipitated by logging and homesteaders who had once attempted to farm, resulting in the transformation of entire landscapes and ecosystems as well as the distribution of wildlife. "I did not find much of the stands of pine... nor anything of the once extensive white cedar swamps. The savannah was enlarged and the swamps and bogs less diversified."[54] While such recovering wetlands rarely replicate fully what he termed a *primeval* state, he understood that even after disruption, native species return through self-regulating processes such as plant succession and diversification, with or without additional human intervention. Conversely, a woodland species such as moose might actually benefit temporarily from human-induced disruption such as from logging or fire, as they might produce intermediate stages of forest succession favoring plants that provide optimal forage such as birch, maple, and willow. A subsequent spike in the number of moose, however, could cause severe overbrowsing and a crash in the population.

Habitat degradation can trigger a vicious cycle, such as "when overabundance of deer for the supporting capacity of the range not only results in their starvation but also in extremely serious damage to that range."[55] Consider the case of Superior National Forest, emblematic of habitat management in the lake states of the Upper Midwest generally. "The most influential factor in reducing the deer from peak populations in the 1930s," Errington asserts, was not in fact predation by wolves as was widely assumed, but "range deterioration, accompanied by starvation."[56] Following sustained overbrowsing, he concluded, "the real supporting capacity of the deer range can go down and down."[57] A prey species might eventually disappear completely from sections of its previous range as succession progresses, conditions change, and resulting habitat no longer proves conducive to survival. In the worst-case scenario, he concluded, "overabundant deer could ruin their habitat."[58] In many other parts of the country, in fact, white-tail deer populations had suffered a fate very similar to massive die-offs in the Midwest; Errington cites the Rocky Mountains and the Pacific Northwest as well as New England. Once apex predators had been extirpated from these regions, hunting alone often failed to maintain populations of game species such as deer at sustainable levels. By contrast, regions with viable populations of major native predators such as wolves, in his estimation, exhibited "the most favorable balances between deer and deer ranges.[59] His analysis of human culpability entails an especially close reading of the region's environmental history: how forest succession first triggered by logging and exacerbated by forestry practices such as fire suppression ultimately resulted in a cascade of habitat degradation.

In *Of Wilderness and Wolves*, Errington recounts another classic case of catastrophic wildlife decline due directly to damaged habitat: reindeer introduced to Alaska from Siberia. Herds reached a peak of half a million in the Territory during the 1930s before plummeting twenty years later to just twenty-five thousand. Biologists initially cited a combination of causes, including "careless herding, wolf predation, and excessive human slaughtering"; yet another important variable, Errington explains, was "deterioration of the range brought on by the reindeer themselves" through severe overbrowsing due to overpopulation.[60] This degradation of habitat, exacerbated by human-induced burning of vast tracts, whether deliberate or inadvertent, also affected herds of caribou native to Alaska where the ranges of these two species overlapped. Contemplating such lasting damage to Arctic habitat and wildlife inspired some of Errington's most strident writing: "Upsetting the whole tundra relationship between animals, plants, soils, and climate cannot be defended in terms of progress or advance of civilization."[61] This warning, formulated by Errington in the 1950s, seems especially prescient today in light of the pipeline from Prudhoe Bay, offshore drilling in Arctic waters, and above all global warming.

Errington realized that ancient interactions between predator and prey—as well as self-regulating processes between each species and its habitat—are extraordinarily intricate and complex: a grand matrix of relationships. "Life has had a long time to make adjustments," he surmised, "animal to animals and to plants, plants to plants and to animals, animals and plants to their physical environments and to the stresses of emergencies," adding wryly, "long before any modern biologist tried to analyze population mechanisms."[62] In more settled areas a host of agricultural activities tend to compound landscape change and hasten the degradation of habitat in the process, whether "trampling or heavy grazing by livestock, farmers hauling in corn shocks or cutting brush along roadsides or removing woodpiles or plowing fields and draining ponds."[63] Accordingly, his guiding principle for preserving natural areas "where naturalness itself constitutes the supreme value" became minimizing interference through invasive management practices while allowing areas previously drained and developed to revert to their natural state.[64] Modest approaches to ecological restoration, he believed, such as the reintroduction of species historically endemic to a place, might complement and accelerate naturally occurring biological processes.

Errington recognized that safeguarding global bird migrations presented particularly vexing questions for conservationists due to the vast and complex networks of ecological niches essential to sustaining each species. He chronicled the rise in species biodiversity as migratory birds arrived in the Upper Midwest, both those nesting locally and others passing through on flyways en route to breeding grounds farther north. *Of Men and Marshes*, for instance, which follows the progression of seasons, begins with great migrations on midwestern flyways in spring. With the eye of a practiced field biologist, he cataloged the remarkable array of waterfowl that were still returning to the region's wetlands at midcentury, including "white or gray herring gulls, flocks of pelicans or cormorants, or herons, grebes, coots, rails, terns, shorebirds, swallows, blackbirds," as well as baldpates, bluebills, buffleheads, coots, goldeneyes, loons, mallards, mergansers, osprey, ringnecks, shovelers, teal, and whistling swans.[65] Especially during periods of peak migration, a single site might be simultaneously visited by a host of different species. For instance, in "An Iowa Marsh," his depiction of Goose Lake, a wetland he visited repeatedly over thirty years, celebrates its thriving populations as well as a species biodiversity that he terms "a superlative abundance and variety of birdlife."[66] While the sheer variety of migratory bird species seems remarkable in itself, their staggering numbers are astonishing.

As a close observer of population dynamics, Errington noted how historic migrations can change due to many influences, ranging from weather patterns to hunting pressure. Yet anomalies abound: A new species of waterfowl altogether may arrive, for example, flying alongside other types of ducks, such as "the comparatively scarce redheads and canvasbacks, sometimes in a place where they have not been seen for several years."[67] And spikes routinely occur in individual species, with thousands visiting a particular locale instead of the mere hundreds the year before. Such fluctuations are only one of the vagaries of migration. Ongoing changes in range are often so fluid and dramatic that he deemed it sometimes "difficult to judge what may be a stray and what may be a successful pioneer."[68]

Errington had warned that by midcentury iconic migratory species such as cranes were already becoming extirpated in some places and threatened in others. He documented reductions in range that had occurred over his own lifetime: "When I was young, the sandhill cranes, large relatives of the rails and coots, were of the north-central prairies," but now, "away from the scattered areas in the northern Lake States where a few cranes persist, one has to go westward in our region to see real free-living cranes or to hear their bugling."[69] The grandeur of the great blue heron inspired in his imagination intimations of deep time: "Their flight is ungainly yet has its grace. One watching it can imagine himself watching the flying reptiles of a younger earth."[70] As naturalist Peter Matthiessen observes, "the first cranes appeared some 50 million years ago, and the sandhill crane... is the most ancient bird species on earth.... They symbolize, like no other living things, the disappearing expanse of clean water, earth, and air upon which their species—and ours, too, though we learn it very late—must ultimately depend for survival."[71]

Errington recognized that such great flocks might one day simply disappear in the wake of pervasive landscape change, especially when several essential ecosystems are being compromised at once along migratory flyways. He reasoned that while historical changes in range reflect dynamic and ongoing ecological processes, age-old routes could be jeopardized entirely by destruction of the vital habitat that provides forage for migrating flocks. Draining wetlands for conversion to cropland, he understood, interferes with and ultimately disrupts even major migrations between breeding and wintering grounds, reducing the range of species whose survival depends on specific types of habitat. Such migration routes had evolved over many millennia, often in response to the broad sweep of geological forces. He realized that some migrations actually predate present-day topography by perhaps *millions* of years. In addition to the daunting distances crossed, these great migrations reflect enduring ecological relationships.

Environmental writers today continue to document the declines in migratory species that Errington had anticipated, and which can now be seen as the harbinger of a growing global crisis exacerbated by climate change. As literary naturalist Scott Weidensaul concludes, "Migratory birds pose one of the stickiest conservation challenges in the world. In the past programs focused on saving breeding areas, but experts now realize they must also save wintering grounds and migratory stopovers."[72] As he points out, "wherever a migrant bird goes, it must fit itself into the environment. During migration, it may pass through dozens of different habitats and ecosystems," often requiring it to adjust to the vagaries of weather accompanying seasonal change and "arrive just as local food resources are near their peak."[73] Moreover, migrating birds "must contend with the radical alteration of the continent's middle latitudes—the fragmentation of once-contiguous hardwood and conifer forests into increasingly isolated islands... [while] in many parts of the Great Plains, more than 98 percent of the original prairie has been converted to crops."[74] Accordingly, "the biggest threats to migratory birds... arise from habitat loss and the wholesale environmental changes we have imposed on the natural world."[75]

Both migrations and ranges have been altered by environmental changes, including habitat loss attributable to human agency and greatly compounded today by the effects of global warming. Changes to ecosystems precipitated by natural causes, including glaciation, generally (though not always) proceed at a relatively modest pace, allowing many species to adapt to changing conditions. However, in recent centuries human-induced fragmentation and outright destruction of habitat have occurred more rapidly and on an unprecedented scale. Scientists are increasingly sounding the alarm. As the rates of landscape and climate change outstrip self-regulating biological processes of succession and adaptation, the prospect of mass extinctions grows. The most daunting scenario is the collapse of not just populations of individual species but entire ecological communities, precipitating a "fundamental reordering of the natural world in the course of the next century."[76]

For Errington, cranes—like the wolf—represent nothing less than "wilderness incarnate," precious and imperiled: "In the fullness of geological time, the last crane will trumpet its farewell and spiral skyward."[77] Indeed, many migrations above the Great Plains had been sharply curtailed by the end of his lifetime due to human disturbance along flyways. The potential extinction of such a majestic species seemed to him unspeakably tragic. Describing the grand rookeries of birds that continue to nest in vast colonies, he marveled at the antiquity of ecological niches that such species still inhabit, declaring "theirs is still the dignity of the millions of years."[78] To anticipate their absence, then as now, is an act of witness grounded in an awareness of environmental history. By acknowledging what is both ancient and indigenous, he ascribed a unique value to preserving wildlife for its own sake.

Of Men and Marshes closes on an especially moving and meditative note. Beyond his perspective as a consummate naturalist who eventually published some two hundred studies of thirty-five different species of mammals and birds, he also indulged in lyrical introspection, what might be seen as his "pastoral turn." Having championed the cause of conservation throughout the volume, he turned his gaze toward the heavens. The grandeur of the night sky inspired a reverence for nature that is nowhere more pronounced than when he is speaking of his beloved marshes: "[When] a person thinks of the inconceivable millions of stars, each so much greater than our whole earth... a feeling of humility before the Order that exists in the universe is always appropriate.... The lessons as well as the beauties of marshes await the perceptive, as do the lessons and beauties of the skies, of the seas, of the mountains, and of the other places remaining where man can still reflect upon lessons and beauties that are not of human making."[79] This spiritual impulse underlies his convictions as a conservationist and, above all, pleas for preserving, protecting, and restoring wetlands and marshes.

Many of Errington's greatest insights as an ecologist were inspired by a keen appreciation for environmental history. Observing flocks gather in autumn anticipating migration, he felt moved not only to describe the bounty before him, but to reminisce about the natural abundance of landscapes he had wandered in his youth: "Their wavy lines across the horizon reawaken my memories of virgin Dakota prairies. I think again of unplowed hills and lowlands covered with native grasses and blazing stars and prairie clovers and vetches and sunflowers and asters and compass plants and cone flowers. I think of withering arrowhead leaves over the waters.... My memory-pictures of the 'old days,' when summer transformed into fall, are of countrysides having greater natural riches than those I now find closer to home."[80] The elevated, lyrical diction of this passage suggests a profound emotional valence and his love for the places on which the enduring matrix of wild nature inevitably depends. Contemplating the depth of his own visceral response to the sheer abundance of immense flocks migrating, he speculated that such feelings are primal and widely shared: "Human responsiveness to the sights and sounds of great numbers of these birds moving northward must be due to more than relief that winter is over, or to anything one might call romantic appeal, or to mere symbolism...something basic in many of us, something identifiable in part with love of what is living and free."[81]

Based on extensive experience in the field both as a sportsman and a scientist, Errington concluded that marshes constitute a unique form of "wilderness" and often possess extraordinary levels of biological complexity. His perspective was that of a pioneering ecologist concerned with what today would be called the preservation of species biodiversity: "Civilized man could show some civilization," he wrote, "by preserving in as natural condition as possible representative types of different plant and animal communities before they are lost."[82] Foremost among the reasons for preserving wetland ecosystems is to "safeguard areas where ancient interrelationships of plants, animals, soils, and climates may be studied."[83] Above all, he was certain that "wild things staying wild" was a value worth preserving in its own right.

NOTES

1. The region at that time has been hailed as "the cultural hearth of wildlife ecology and of American ecology in general" (Kohler, *All Creatures*, 227).
2. Errington, *Of Wilderness and Wolves*, 210.
3. Sivils, "Paul L. Errington," xxii.
4. Errington, *Of Wilderness and Wolves*, 159.
5. Errington, *Red Gods Call*, 145.
6. Errington, *Of Wilderness and Wolves*, 164.
7. Ibid., 65.
8. Errington, *Of Men and Marshes*, 2.
9. Ibid., 15.
10. Errington, *Of Wilderness and Wolves*, 162.
11. Errington, *Question of Values*, 107.
12. Ibid., 132.
13. Errington, *Red Gods Call*, 122.
14. Errington, *Of Men and Marshes*, 31.
15. Errington, *Red Gods Call*, 118.
16. Errington, *Of Wilderness and Wolves*, 33.
17. Ibid., 149.
18. Sivils, "Paul L. Errington," xxii.
19. Errington, *Of Wilderness and Wolves*, 164.
20. When trying to place his final manuscript for publication with commercial presses, Errington convinced editors at Macmillan that the intended audience for *Marshes* was the same one that had made Leopold's *Sand County Almanac* a success.
21. See Kohler's "Paul Errington, Aldo Leopold and Wildlife Ecology: Residential Science" for a detailed examination of parallels between their backgrounds, educations, careers, and theories as well as distinctions such as "their different experiences as trapper and forester in the different landscapes of the Southwest and the North" (246).
22. Pritchard et al., "Landscape," 1413.
23. Errington, *Of Wilderness and Wolves*, 15, 40.
24. Sivils, "Paul L. Errington," 6.
25. Ibid., 3–4.
26. Ibid., 31.
27. Errington, *Question of Values*, 107.
28. Sivils, "Paul L. Errington," xii–xiv.
29. Errington, *Question of Values*, 108.
30. Errington, *Red Gods Call*, 165.
31. Pritchard et al., "Landscape," 1412.
32. Kohler, *All Creatures*, 40.
33. Errington, *Red Gods Call*, 47.
34. Ibid., 22.
35. Errington, *Of Men and Marshes*, 10.
36. Errington, *Red Gods Call*, 116.
37. Errington, *Of Wilderness and Wolves*, 35.
38. Ibid., 2.
39. Kohler, "Paul Errington," 243.

40. Errington, *Of Wilderness and Wolves*, 27.
41. Kohler, "Paul Errington," 222.
42. Errington, *Of Men and Marshes*, 141.
43. Indiana Department of Natural Resources, *Prophetstown State Park Interpretive Plan*.
44. Errington, *Of Men and Marshes*, 28–29.
45. Errington, *Of Wilderness and Wolves*, 124.
46. Pritchard et al., "Landscape," 1415.
47. Errington, *Of Wilderness and Wolves*, 195.
48. Ibid., 152–53.
49. Ibid., 154.
50. Ibid., 69.
51. Ibid., 153.
52. Ibid., 66.
53. Errington, *Red Gods Call*, 97.
54. Ibid., 98.
55. Errington, *Of Wilderness and Wolves*, 103.
56. Ibid., 96.
57. Ibid., 97.
58. Ibid., 113.
59. Ibid., 100.
60. Ibid., 84.
61. Ibid., 92.
62. Ibid., 82.
63. Ibid.
64. Errington, *Question of Values*, 147.
65. Errington, *Of Men and Marshes*, 9.
66. Errington, *Question of Values*, 34.
67. Errington, *Of Men and Marshes*, 13.
68. Ibid., 33.
69. Ibid., 18.
70. Ibid., 16.
71. Matthiessen, "Last Cranes of Siberia," 276.
72. Weidensaul, *Living on the Wind*, 22.
73. Ibid.
74. Ibid., 27.
75. Ibid., x–xi.
76. Ibid., 368.
77. Errington, *Question of Values*, 101.
78. Errington, *Of Men and Marshes*, 86.
79. Ibid., 150.
80. Ibid., 33.
81. Ibid., 12.
82. Ibid., 141.
83. Ibid.

4

The Wilds That Gave Us Birth

Scott Russell Sanders (b. 1945)

The rich diversity of the world's cultures reflects
the corresponding diversity in the wilds
that gave them birth.

—Aldo Leopold, *A Sand County Almanac*

In reading landscapes for their environmental history, contemporary midwestern literary naturalists advance the legacy of pioneering ecologists, particularly Aldo Leopold, who grasped the fundamental ecological precept that preserving the diversity of indigenous species helps to ensure the stability and resilience of a biotic community. Scott Russell Sanders is well aware that he carries the environmental torch passed down in our national literature. Contemplating the roots of environmental ethics, he suggests that "the greatest theme in American literature is the search for right relations between humankind and nature, between civilization and wildness."[1] In *A Conservationist Manifesto*, he acknowledges this heritage, naming precursors that have informed his thinking and who inspire him still. "In America, one can trace a lineage of dissident souls—John Muir, Aldo Leopold, Anna and Harland Hubbard, Thomas Merton, Edward Abbey, Wendell Berry, Annie Dillard, and a great many others—who found in Thoreau if not the inspiration then at least the confirmation for their own efforts at rethinking the meaning and conduct of life."[2] Sanders's writings fit squarely within this tradition. "The roots of conservation go deep in America back through such visionaries," he declares in the preface.[3] Moreover, at his most lyrical moments, Sanders clearly echoes the Transcendentalist sensibilities of Emerson and Thoreau. "I belong to this lineage of writers inspired by Thoreau," he adds, "however humble my place may be."[4] Above all, he reminds us, "this lineage continues to inspire those of us who seek to articulate a conservation ethic for our own time."[5] He acknowledges, however, that "our ethical dilemma is quite different from the one Thoreau faced."[6]

Sanders's narratives of environmental history encapsulate the centuries of landscape change *represented* in midwestern literature: forests logged off, wetlands drained, and prairie sod plowed under. Like other writers in the region bearing witness to ecological decline, he laments that it has resulted in the erasure of entire ecosystems, squandering their legacy of biological abundance. He recounts the litany of species completely extirpated from the region, including keystone predators: "Bison, bears, lions, wolves, and passenger pigeons, Carolina parakeets, and countless other animals dwelt here in astounding numbers. This original abundance thousands of years in the making, we have all but used up in two centuries."[7] This chapter examines how his perceptions of landscape change awakened a keen appreciation for environmental history, culminating in a vision that extends Leopold's land ethic in two significant ways: considering the dire state of the global environment today, and positing an ethical response informed by the traditions of indigenous peoples.

Cultural memory of landscape change—what might be termed our collective *ecological memory*—encompasses the cumulative alteration of self-perpetuating natural ecosystems over centuries if not *millennia* of human use. To put the transformation of the midwestern landscape into historical perspective, Sanders asks us to consider that "in 1800, the grasslands that we glimpse now in tiny scraps would have stretched westward to the Great Plains; the glacial wetlands that we've almost entirely drained would have stretched north to the Great Lakes and beyond, up to the ice-gouged vicinity of Hudson Bay; the hardwood forest that we've reduced to rare pockets of big trees would have stretched eastward all the way to the Atlantic and south to the Appalachian Mountains."[8] Here he anticipates subsequent literary naturalists such as Paul Gruchow and Elizabeth Dodd, who also echo Leopold by recounting the geological processes of glaciation that shaped the topography of the Midwest, creating extensive wetlands at the end of the last ice age.

Perhaps Sanders's most sweeping evocation of environmental history appears in his foreword to a recent reprinting of Gruchow's seminal volume of essays *Journal of a Prairie Year*. He attributes Gruchow's view of habitat preservation and stewardship by restoration to his awareness of landscape change in the region accelerated by centuries of human use.

> At the beginning of the nineteenth century, travelers setting out from the southern lip of Lake Michigan, where Chicago sprawls today, could have rolled westward to the Rocky Mountains through an uninterrupted sea of grass…waded through Indian grass, switch-grass, coneflowers, sunflowers, blazing stars, goldenrods, wild indigos, and hundreds of other species of plants. This swath of tall-grass prairie stretched north and south in a sinuous belt more than a thousand miles long.… And everywhere they would have found an abundance of game, for these grasslands nourished immense herds of bison, as well as antelope, deer, rabbits, geese, and ducks.[9]

Consulting the historical record, Sanders establishes a kind of ecological baseline against which to measure ongoing landscape change. In "The Force of Moving Water," for example, he recounts the natural abundance once found along the Ohio River and expresses a deep-seated desire to have witnessed the "primal country" of the Midwest prior to European settlement, "one that preceded all maps."[10] At the time of initial exploration, the watershed was renowned for its abundant wildlife. Among the earliest British explorers to assess the "raw abundance" of the Ohio River Valley was Christopher Gist, who visited the region in 1750–51, and depicted a verdant landscape ripe for settlement and cultivation, with plenty of game to sustain the first wave of pioneers: "beautiful natural Meadows, covered with Wild Rye, blue Grass and Clover, and abounds with Turkeys, Deer, Elks, and most Sort of Game particularly Buffaloes."[11]

Accounts by other early adventurers such as George Croghan in 1765 describe a profusion of wild game to hunt—notably bears, deer, and buffalo—providing easy pickings for frontier hunters. The banks were then a verdant woodland brimming with biodiversity, the understory laced with native wild grapevines. By 1770, when George Washington was surveying and acquiring property along the Ohio, he exclaimed at its sheer bounty while paddling the river: "this country abounds in buffaloes and wild game of all kinds; as also in all kinds of wild fowl," including ducks, geese, and swans, as well as "fine fish."[12] During the first decades of the nineteenth century, Sanders reports, the region still sustained impressive populations of wildlife, including "panthers and wildcats, fox and lynx, otters and minks, herds of elk, packs of wolves, whooping cranes by the flock, ducks and geese and partridges and pheasants by the skyful."[13] In fact, Sanders points out, pockets of such abundance endured throughout the Ohio's watershed well into the nineteenth century, with bears, beaver, deer, otters, muskrats, and squirrels common along the banks, and fish of over a hundred kinds still plentiful in the river. Artist and naturalist John James Audubon chronicled much the same ample supply of game in the region during the first half of the century, noting especially bountiful species of birds such as grouse and teal as well as the iconic American turkey.[14] However, he had already begun to lament a pattern of rapid environmental decline precipitated by initial white settlement and subsequent development in many parts of North America, including the Midwest. For Sanders, accounts such as Audubon's, and those of his predecessors, provide a frame of reference for appreciating the historic vitality as well as the ongoing vulnerability of the region's habitat and wildlife.

In aggregate, Sanders concludes, "the primeval forest which in 1800 had covered some twenty million acres of Indiana territory... by 1865 had already become rare. In two thirds of a century, nearly all the forest had been cut, the prairies plowed, the swamps drained."[15] Deforestation proceeded at an alarming pace in that era, with a staggering seven thousand acres cleared a day. An estimated *half a billion* trees were felled by hand. As the nineteenth century progressed, exploitation of such natural resources increasingly took its toll, as Sanders recounts in *Staying Put: Making a Home in a Restless World*, where he elaborates on the impact of such settlement and development along the Ohio River:

> The pattern of habitat destruction and relentless hunting has been repeated for species after species. Bison, whose wide traces offered routes through the woods for the first roads, were still counted in hundreds at salt licks near the river on the eve of the Revolution, and yet within a few years they had been killed off. The lynx, wildcat, panther, elk, otter, bear, and wolf disappeared from the region; the green parakeet vanished altogether. The whooping crane dwindled almost to extinction, and so did the bald eagles. By 1900, coal slurry and oil slicks from wells and mines on the upper river had killed off untold species of fish.[16]

Today only scattered parcels of native woodland and prairie remain preserved following "the annihilation of the forests and disappearance of wildlife... the squandering within a few generations of an unforeseen, unearned bounty."[17] Indeed, subsequent centuries of development have resulted in the epic diminishment of natural ecosystems throughout much of

North America. Viewed from this broader historical perspective, he concludes, the fragmentation of contiguous natural habitat has reached such an extreme that "after living for thousands of generations in tiny settlements surrounded by wilderness, in the past half century humans have reduced wilderness to a scattering of remnants surrounded by our own settlements."[18] The fertile grasslands of the midwestern "Heartland," he explains, have been no exception: "By the mid-twentieth century... the tallgrass prairies that had once nurtured big bluestem and bison had become little more than a memory. All but a few remnants of these millions of acres had been plowed or paved; creeks had been straightened into ditches; potholes had been drained; vast herds and flocks had dwindled or entirely disappeared."[19]

Sanders distinguishes his own perspective—and the imperatives of contemporary conservationists generally—from those of the nineteenth century. Thoreau, he observes, "did not practice simplicity because he was worried about damaging the planet or depriving future generations."[20] In our own time, by contrast, the environmental consequences have grown ever more profound and, as Sanders sees it, now approach the absolute: "The dynamic equilibrium of the biosphere... is the single most important factor in the continued flourishing of life on Earth. Any human activity that disturbs this dynamic equilibrium... is a threat not only to humankind but to every other species."[21] Accordingly, he warns of the peril of overharvesting that outstrips natural processes that can replenish sustainable populations:

> Up to a point, trees may be harvested from a forest, crops may be harvested from the fields, fish may be harvested from the sea, and the regenerative power of nature will replace what has been taken away. If pushed far beyond that point, however, forests give way to deserts, as in North Africa; soils become sterile, as in much of the Middle East; and fish stocks collapse, as has happened recently to dozens of species, such as cod, that were once a mainstay of the human diet. No form of consumption is sustainable, therefore, if it exceeds the capacity of a natural system to replenish itself.[22]

"Unlike Thoreau and his contemporaries," Sanders concludes, "we now face problems that are global in scale."[23] Moreover, he declares, "the scale of devastation caused by human activity is unprecedented, and it is accelerating.... The spreading of deserts, clear-cutting of forests, extinction of species, poisoning of air and water and soil."[24] The US Department of Agriculture calculates that some 2.2 million acres of farms, forests, wetlands, and prairies are lost to development every single year. He warns that the cumulative impact of ecological destruction may cause irreparable harm to the environment, since such extensive disruption often seems virtually irreversible: "Once a wilderness is cut up by roads, oil-drilling platforms, landing strips and toxic dumps, it will never again be wilderness, at least within many human generations."[25]

In the continent's heartland Sanders sees a paradigmatic example of extensive environmental decline to a biome exacerbated by the pace of human development. "The dream of dominating the wilderness," he contends, "is the story of the whole continent in miniature... played out here in the Heartland with fiercer energy and more bitter disillusionment.... We have labored to control nature with a thoroughness and zeal unmatched by any other region."[26] Moreover, he observes, "of all the regions in America, the Midwest is most easily—if superficially—subdued, and therefore the one where the failure of our efforts, the waste of riches, the betrayal of promise is most painfully evident. If literature of the Midwest began as the story of arrivals and departures, it has evolved into a literature of loss."[27] He laments how the fertility of the grasslands that had taken millennia to create could be so severely compromised in less than two centuries through a combination of soil erosion and the depletion of nutrients. The loss of fertility has led to a dependence on petrochemicals and, more recently, genetic engineering to increase yields at the expense of further environmental degradation. He concludes: "a legacy of fertility built up on the prairie over thousands of years had been squandered."[28]

Yet even when land use is discussed in economic terms, the long-term environmental costs of development are typically left out of the calculations. Consider, for example, the prevailing model for agribusiness in the Midwest, where the fundamental soil-plant-animal-human food chain posited by Leopold is stripped of both its biological and spiritual integrity. As Sanders describes, "the deep topsoil is devoted mainly to soybeans and corn and is liberally sprayed with pesticides and herbicides."[29] This monocultural regime steeped in petrochemical compounds wreaks environment havoc on a great many of the region's watersheds. Moreover, factory practices of modern livestock operations compound the water pollution created by cropland runoff tainted with a toxic mix of fertilizers, pesticides, and herbicides. Enormous quantities of hog manure, laced with contaminants ranging from antibiotic residues to heavy metals, are applied to the same fields. Dangerous concentrations build up in the soil and over time pollute waterways throughout the region. Sanders describes the case of the White River in Indiana: "The resulting stew of toxins has made it dangerous to drink straight from the river, swim in the river, or eat fish drawn from the river."[30] The vast dead zone at the mouth of the Mississippi, he points out, represents another direct and especially devastating consequence: "Agricultural runoff flushed down the Mississippi River has extinguished nearly all life in an area of the Gulf of Mexico that ranges from 5,000 to 8,000 square miles."[31] To make matters worse, governmental entities charged with safeguarding environmental quality do little more than preside over the ongoing degradation: "the word [environmental] *management* is a misnomer here," he points out, since, ironically, "at best our state and federal agencies are monitoring its decline."[32]

Sanders characterizes present-day agricultural ventures in the Midwest as the culmination of a long-held vision of European settlers in the Americas, who routinely moved on after denuding entire landscapes of verdant ecosystems. The sprawling industrial farms of today's agribusiness, he suggests, perpetuate that sad legacy; they are "the latest expression of our drive to simplify the land."[33] Yet he is well aware that similar declensionist narratives have unfolded repeatedly over centuries and millennia in a variety of geographic contexts, as decried in the mid-nineteenth century by visionaries such as George Perkins Marsh, as well as writers in our own time, ranging from Wendell Berry and Peter Matthiessen to David Quammen and Gary Snyder. Emblematic examples include the early deforestation of much of Asia, including India and China; the Industrial Revolution radiating from Europe around the globe; the Dust Bowl erosion in the United States; and the ongoing deforestation in southeast Alaska and the Pacific Northwest, as well as in tropical rainforests. Sanders characterizes with dismay the severity and cumulative extent of damage to the environment: "Pollution from rivers has created dead zones in the oceans.... Through the use of heavy machinery in farming, we have lost much of our topsoil to erosion, and through the application of poisons we have reduced the fertility of the soil that remains. Through the clearing of forests, especially near the equator, we have enlarged the reach of deserts. By draining wetlands [and] paving fields...we are driving to extinction countless other species."[34]

Throughout many volumes of essay and memoir, Sanders returns to the theme of landscape change: particularly the conversion of woodland, wetland, and prairie for crops and livestock, a process culminating in the practices of modern industrial farming. This cycle of habitat loss, wildlife decline, and local extinction is emblematic of several centuries of ecological degradation on the American frontier. In addition, the erasure of natural habitat by extractive industries such as mining and logging has diminished the natural abundance of North America to a fraction of its original, pre-contact extent—an environmental travesty. Sanders laments that only "remnants . . . of land that have survived in something like their pre-settlement condition have been left undisturbed, and that . . . less than 1 percent of the territory that became Indiana remains in our day relatively pristine, unaltered by saws and bulldozers and plows."[35] In our time only fragments survive: "some five hundred acres, never plowed nor paved is the largest remnant of grasslands that once covered a few million acres of Indiana." Ironically, today this preserve borders an industrial wasteland.[36] Such narratives of landscape decline underlie Sanders's calls for environmental advocacy and form the basis for reenvisioning ethical relations to nature.

How exactly did Sanders arrive at an ethical stance toward our relationship with nature? As with many other ecologically minded midwestern literary naturalists of the last century, notably Leopold, Errington, and Gruchow, his perspective began with childhood experiences of engagement with nature and personal acquaintance with the extent of landscape change. In his memoir *The Country of Language*, Sanders recounts some of his earliest experiences in the wild as he "learned to read the woods and fields."[37] He describes being introduced to the art of tracking by his father, an avid hunter: "scanning for animal tracks. Here was another alphabet for me to learn. . . . As I learned to recognize animal tracks and tree bark and leaf shapes from my father, so I learned the alphabets of flowers and vegetables and fruits from my mother, who was always a gardener."[38] And in the memoir *A Private History of Awe*, Sanders vividly depicts these early experiences that kindled his lifelong passion for both nature and language: "On our walks, I begged to know the names of garden plants, crops in the fields, trees in the woods and so I learned peanuts and okra, sorghum and burley, ironwood and sweet gum. After I wore out my father's patience, I would go ask my mother the names of flowers and spices. *Zinnia. Peony. Oregano. Basil.*"[39]

Sanders's perception of environmental history was initially shaped by places he knew as a youth, including twenty-one thousand acres of former farmland in Ohio surrounding a munitions plant where his father worked.

> Scattered through those thousands of acres were hundreds of foundation holes from farmhouses and barns that had been bulldozed in 1940 for the building of the Arsenal. There were abandoned orchards, beds of asparagus and rhubarb, clumps of perennial flowers that kept pumping out blossoms among the weeds, and family cemeteries growing up with brush.... Muskrats paddled through the scum leaving black trails. Beavers built dams in the creeks. Foxes hunted mice in the fields. High in the tallest trees, eagles made nests out of sticks. Deer meandered everywhere.[40]

Wildlife flourished so well that the US Army eventually hired a trapper to eradicate what were considered unwanted vermin—including coyotes, foxes, beavers, muskrats, and mink—adding to the litany of species eliminated there in an earlier era. As the trapper puts it: "The old-timers got every last lion and bear and buffalo before we ever came along."[41]

Sanders asserts that memories of places from childhood exert a lifelong influence on our relationship with nature, and perhaps nothing is more powerful than witnessing the destruction of landscapes we once cherished. "No fate," he declares, "could be more ordinary" than to face the desecration of such hallowed ground, those very places that have awakened our personal appreciation of nature.[42] Consider how his essay "After the Flood" depicts landscape change to his childhood home in rural Ohio. He vividly recounts the dilapidated state of the place when his family arrived: "It was a failed farm, Dad said, which made it cheap enough for us to buy. The fences were mostly down. Plows, a harrow, a manure spreader, an iron-wheeled tractor, and other old equipment rusted in a ravine at the edge of the woods. The pigpen was a warped box on skids."[43] Returning there as an adult, he attempted to square his memories and emotions with a place that had been irrevocably altered when the neighboring Mahoning Valley was flooded.

His family's farm had been abandoned when the river was dammed, an event that left an indelible mark on Sanders's imagination—and a deepened appreciation for the ecological processes that commence when any developed tract of land is left to revert to a feral state. "A dam was built," he explains, "the river died, and water backed up over most of the land I knew"; nonetheless, even after many years' absence, he tells us, this remained for him the "place by which to measure every other place."[44] Moreover, he asserts, "One's native ground is the place where, since before you had words for such knowledge, you have known the smells, the seasons, the birds and beasts.... Even if you become intimate with new landscapes, you still bear the impression of that first ground."[45] A particularly lyrical reflection in *The Private History of Awe* expresses an innate desire to perceive nature intimately, to identify with the wild: "A child is more like a forest, gathering every drop of rain or flake of snow, every fallen leaf, the slant of sunlight and glint of moonlight, the fluster and song of birds, the paths worn by deer, the litter of bones and nuts and seed, and whatever the wind delivers, taking it all in, turning everything into new growth."[46] Indeed, much like Errington and Gruchow, Sanders recalls how as a schoolboy he had fantasized about belonging to the wild: "I wanted to go my own way, like a coyote that steals through the woods, sniffing and snooping, pondering things, howling now and again in the safety of darkness, evading all traps."[47] "For as far back as I can remember," he confides wistfully, "I've never doubted that I'm an animal.... I've always understood that my flesh is made from other flesh. On our farm we ate chickens that had scratched in the yard and pheasants that once lay on the kitchen counter wrapped in their shining feathers like a coat of stars."[48]

Revisiting his family's former homestead as an adult, Sanders found that the landscape and farm buildings fixed in his memory were gone. Forest had already begun to return. "The field where I had baled hay," he discovered, "now bristled with a young woods."[49] Trees he had helped plant as a boy were now "fifty feet high, brimming over the woods that used to be our cleared land."[50] He realized that in the time since his departure as a child, "all the while, cedar and sumac and brambles, like the earth's dark fur, [had been] pushing up though my past."[51] A succession of native plant species, undoubtedly mixed with a smattering of invasives and ornamentals that had naturalized, had, in the span of half a human lifetime, begun to establish a renewed woodland community. A neighboring property once belonged to Sanders's high school science teacher, who, much like Stratton-Porter, had cultivated a meadow of wildflowers by gathering seeds from the surrounding countryside, and had followed the Indian custom of burning periodically to forestall its reversion to woodlands. Yet cedar, oak, and staghorn sumac had reclaimed the place completely in the twenty years since her passing, while a dilapidated cottage adjoining the parcel had become entirely enveloped in trees. Here Sanders is practicing exactly the sort of "reading" of landscape that Leopold espoused, an attention to a place's environmental history that includes identifying the type of biome that thrived prior to development and would gradually return in its wake.

As he concludes in the essay "Telling the Holy," "the Mahoning River, long since dammed, still runs in me.... In memory, a forest I have not seen for twenty years still murmurs with the voice of my father naming trees.... I am bound to the earth by a web of stories.... By keeping the stories fresh, I keep the places alive in my imagination."[52] In fact, the grief he experienced on returning to find his family's farm buildings razed, pastures overgrown, and the river valley itself submerged gave rise to an intensified desire to preserve the natural world whenever possible. "This ground was lost," he conceded. "But other ground could be saved, must be saved, in every watershed."[53] Here, Sanders once again exhibits his philosophical kinship with other twentieth-century literary naturalists, espousing environmental ideals and philosophical principles that serve as a foundation for present-day thinking about conservation.

Among pioneering ecologists of the twentieth century, arguably none has had a greater or more direct influence on Sanders than Aldo Leopold. Both explore an alternative paradigm in which conservation is framed in not only pragmatic but ethical terms, and both recognize that the future of the biosphere, and of humanity itself, is predicated on the integrity of the natural world. Sanders contends that "Leopold remains a vital figure for us today because he analyzed the sources of ecological damage with unprecedented clarity."[54] As Leopold had declared, establishing a harmonious relationship with the natural world "is impossible without a visual knowledge of the land's history."[55] Considering both biological processes and human perception of nature allowed Leopold—as well as his contemporaries, such as Sigurd Olson—"to present fundamental ecological principles in a way that would hold the interest of a non-scientific audience, and to inculcate love and respect for natural communities through what he called 'the land ethic.'"[56] In this formulation, any strict disciplinary divide between the sciences and humanities precludes a more nuanced appreciation for ecological processes and their significance for human culture and civilization. Pairing science and ethics in the title essay of *A Conservationist Manifesto*, Sanders calls for just such an interdisciplinary approach: "If we are to foster a culture of conservation, we will need to draw on the wisdom and moral passions of religious communities.... We will also need to draw on a full spectrum of science."[57] For science alone, he reminds us, runs the risk of eclipsing the reverence rooted in the mystical dimension of religion, since "the achievements of science delude many into thinking that we have graduated from nature."[58]

Sanders acknowledges that "Leopold gave us the beginning of this new ethic nearly a half a century ago, in *A Sand County Almanac*, when he described the land itself as a community made of rock, water, soil, plants and animals."[59] He likened the book to "an ark... ferrying an ethical vision through stormy times."[60] As a wildlife ecologist and professional forester, Leopold had been particularly well positioned to consider the environmental consequences of land-use policies and above all the preservation of expansive tracts of relatively wild habitat. The foreword to the highly influential *Almanac*, for instance, combines scientific reasoning with an ethical perspective to address such issues: "That land is a community is the basic concept of ecology," he proclaimed, "but that land is to be loved and respected is an extension of ethics."[61] Leopold also introduced a philosophical and, at times, even spiritual dimension to the discussion of environmental ethics. Writing in the 1940s, he declared in his seminal environmental tract, "There is as yet no ethic dealing with [humanity's] relation to land and the animals and plants which grow upon it."[62] Sanders reminds us, sixty years later, that we *still* "need a story that celebrates the true source of our wellbeing," namely ecological integrity.[63] This has led ecocritic William Barillas to hail him as "the Midwestern writer who has done the most to develop the implications of Leopold's land ethic and aesthetic for the nature essay."[64] Indeed, Sanders's *Manifesto* in many important respects follows directly from a central tenet of "The Land Ethic": "All ethics so far evolved rest upon a single premise: that the individual is a member of a community of interdependent parts.... The land ethic simply enlarges the boundaries of the community to include soils, waters, plants, and animals.... A land ethic of course cannot prevent the alteration, management, and use of these 'resources,' but it does affirm their right to continued existence, and, at least in spots, their continued existence in a natural state."[65]

Leopold's *Almanac* has achieved nearly iconic status, being "often referred to as the bible of the modern environmental movement."[66] Scholars such as Robert Finch credit Leopold's book with no less than "providing the core for modern conservation ethics."[67] Yet as ecocritic Daniel Philippon observes, "*Sand County Almanac* is as much a work of environmental history as it is a work of environmental ethics.... The land ethic emerges from, and is dependent on, a deep sense of one's own environmental history."[68] In fact, Leopold is widely credited with effectively founding several pioneering disciplines, including conservation biology, environmental ethics, and restoration ecology. Environmental philosopher Baird Callicott deems him a "watershed figure," likening his contribution to conservation to those of George Perkins Marsh, John Muir, and Henry David Thoreau.[69] By the 1970s, essayist Sigurd Olson was among the many prominent environmental advocates who hailed the land ethic as the founding principle of modern conservation movements: "There have been many definitions of conservation. [When] Aldo Leopold said 'It is the development of an ecological conscience,'...he implied unless we develop a feeling for the land and an understanding of it, unless we become one with it and recognize love, stewardship, and above all an appreciation of its intangible values, we cannot fully comprehend what conservation means."[70] As Sanders concludes, *Almanac* provided "an ecological standard for judging our actions."[71]

In *A Conservationist Manifesto*, Sanders seeks to reorient contemporary environmental discourse by reintroducing such ethical dimensions, advocating public dialogue about preserving ecological balance based on moral principles. Moreover, *Manifesto* explores the many senses in which human civilization—indeed, our very existence as a species—depends on the ecological balance of the natural world. Sanders has long seen the natural world as text to be read, like Leopold, who had argued that to fully appreciate a place involves recognizing changes to the landscape, especially how human use has repeatedly reshaped wild ecosystems and too often compromised their integrity. In fact, Leopold sought to instill conservationist sensibilities in future ecologists, beginning with the capacity to read landscapes for evidence of ecological processes and their disruption over time. Reading nature as text had been a recurrent trope for Leopold, who likened weeds in a church cemetery to books of history and declared every farmstead a veritable textbook on ecology. He exhorted his graduate students to examine agrarian terrain for evidence of environmental history and to determine whether a forest or grassland biome had preceded settlement. Close reading of nature and place, he felt, would inevitably lead beyond aesthetic appreciation. Such intimate familiarity could engender both deep affection for the wild and a concomitant sense of responsibility: the ethical imperative to preserve ecological integrity, even on the feral margins that remained in the wake of development.

Sanders frames his own environmental creed in terms Leopold would readily have recognized: "It's about claiming kinship. When I say that I belong to this town or watershed or region, I'm declaring membership, as in a family, and I acknowledge my obligation to behave in a way that honors, that protects the whole of which I am a part... [which affirms] bonds of affection and responsibility between person and place."[72] He invokes a passage from *A Sand County Almanac* to characterize the ethical concerns still faced by contemporary conservationists: "We abuse land because we regard it as a commodity belonging to us. When we see the land as a community to which we belong, we may begin to use it with love and respect."[73] He acknowledges human culpability for ongoing deforestation, pollution, erosion, and climate change and the corresponding sense of responsibility for protecting the ecological integrity of the natural world. He expresses this latter sentiment especially forcefully in *Hunting for Hope:* "We bear a solemn obligation to conserve the earth's bounty, for all life. This means we should defend the air and water and soil from pollution and exploitation. It means that we should protect other species and preserve the habitats they rely on."[74] Yet he also offers an alternative, more hopeful vision of environmental restoration. Preserving the integrity of self-sustaining ecosystems with a full complement of native species in our time, he contends, necessitates the restoration of biodiversity.

Sanders argues, moreover, that environmental health is contingent on "dynamic integrity not only in a big wilderness but also in farms and forests that have been cared for through generations."[75] He recognizes that even relatively small parcels of neglected acreage are wellsprings of biodiversity: "Every unsprayed garden and unkempt yard, every meadow, marsh, and woods may become a reservoir for biological possibilities, keeping alive creatures who bear in their genes millions of years' worth of evolutionary discoveries."[76] Even seemingly incidental and fragmentary patches of habitat are miniature ecological preserves, he contends, that can provide a source of spiritual inspiration reminiscent of Muir's celebration of extensive tracts of largely undisturbed wilderness: "Every such refuge may also become a reservoir for spiritual possibilities, keeping alive our connection with the land, reminding us of our origins."[77] In fact, this is why contemporary restoration ecologists routinely collect rare native plants from incidental refugia provided by cemeteries, railway right of ways, and stream embankments fenced off from livestock. Seeds gathered and specimens transplanted from these tiny safe havens may offer the only stay against their extinction.

In *Hunting for Hope*, Sanders cites an encouraging litany of ongoing efforts at environmental stewardship throughout this country and the world:

"returning animals and plants to areas from which they had vanished, reflooding drained wetlands, gathering rare seeds, replanting forest and prairies, cleaning up rivers, helping endangered species and battered lands to recover."[78] In this respect, he is heir to Indiana's Gene Stratton-Porter, whose *Girl of the Limberlost* was among the most widely read novels of its time for its evocation of primordial wetlands and their ecological bounty as well as the redemptive power of nature. She is one of several preeminent writers from the region to portray the power of the wilds to restore both physical and emotional well-being. As Sanders explains, "The Limberlost she relished was untamed.... She celebrated the fecundity and beauty of the Limberlost, which appears in her books and as a kind of Eden."[79] Later in life Stratton-Porter set about restoring endangered native species by transplanting wildflowers by the thousands to her property at Limberlost North on Sylvan Lake. In "Telling the Holy," Sanders recounts a personal pilgrimage he made to one such site, Loblolly Marsh Wetland Preserve, to witness a modern-day restoration of a 428-acre fragment of the original Limberlost Swamp, "a remnant of the vast, magnificent, vanished wetland" in Indiana's Jay County.[80] His account of current restoration efforts and evocation of increasing biodiversity there reads like a Whitmanesque catalogue of native species:

> Stems of last year's foxtail, switchgrass, Indian grass, big bluestem and little bluestem wave on the uplands.... These grasses were planted, along with a number of wildflowers—blue flag iris, purple coneflower, wild bergamot, indigo—but other plants have come back on their own, including milkweed and hemp dogbane.... White swaths of boneset are blooming now, along with bold yellow wands of golden ragwort and demure white clusters of daisy fleabane.... Broad-leafed and narrow-leafed cattails have found their way here to crowd these pools, and so have rushes, sedges, plantain, smartweed, cottonwood, and willow.[81]

Indeed, this thriving vestige of a verdant wetland ecosystem now brims with over 350 species of native plants, including black oak, blackberry, blazing star, bluestem, cottonwood, leadplant, rattlesnake master, sunflower, white oak, and wild indigo. Though a good number of these plants have been deliberately reintroduced, others have reappeared of their own accord: "Seeds from some of these native species may have lain dormant in the muck for the past hundred years, waiting for the corn to go away and the water to return.... The marsh is rousing, as if waking from a long sleep," he concludes. "Ancient rhythms...are slowly returning to this long-used, much-loved, and richly imagined ground."[82]

While Sanders's *Manifesto* represents an extension of Leopold's land ethic, it also breaks new ground, such as when contemplating Native American perspectives on the natural world. He believes unequivocally that preservation of traditional cultures, languages, and lifeways is part and parcel of protecting ecological integrity. The corollary is to incorporate concerns for cultural diversity and social justice within the discourse of environmental ethics. In fact, he deems the two to be inseparable. In this formulation, diversity is to human well-being what biodiversity is to ecological communities. Moreover, the two are in a sense seen to be reciprocal. "Just as DNA stores within our bodies a deep evolutionary inheritance," he observes, "so language stores outside our bodies the cumulative discoveries of our species."[83] Genetic diversity represents the biological legacy of evolutionary processes; human diversity reflects the accumulated history of inhabiting a wide variety of landscapes in myriad ways. Biological and cultural diversity are evidently intertwined by DNA's double helix, as if wound on a single spindle. Yet Sanders's inclusive vision is decidedly consistent with Leopold's: "Our effort to honor human differences cannot succeed apart from our effort to honor the buzzing, blooming, bewildering variety of life on earth."[84] In what ways, then, might we extend the celebrated land ethic today, particularly given the cultural legacy of the continent's indigenous peoples?

In his essay "Telling the Holy," Sanders contemplates the extent to which indigenous cultural memory perished as tribes were displaced and white settlers laid claim to land in their wake:

> We live in a land that has been known, remembered, spoken of with reverence and joy for thousands of years.... In my own region of the Ohio Valley, there are few traces left of the aboriginal way. As the Shawnee, Miami, and other tribes were driven out, by arms or treaty, we lost the benefit of their long-evolved knowledge of the animals, the plants, the seasons, the soil itself. We lost nearly all of their stories and songs.[85]

This pattern was tragically repeated across the continent as a "hundred other tribes were herded onto reservations far from the graves of their ancestors."[86] Sanders asserts that the languages, stories, and lifeways of traditional subsistence cultures represent an essential heritage for humanity as a whole. "What we call landscape," Sanders reminds us, "is a stretch of earth overlaid with memory, expectation, and thought," and in saying so he suggests the imperatives of stewardship and restoration.[87] The relationships of indigenous peoples to the landscapes of North America, he recognizes, are primordial ones: "We glimpse a past reaching back more than a thousand years in the architecture and lifeways of Hopi pueblos or Tlingit villages, and we sense an even deeper past among the ancient earthworks of the Upper Mississippi and its tributaries, at Cahokia in Illinois or the Serpent Mound in Ohio."[88] Ultimately, he contends, "in seeking a way of life that is durable we have much to learn from those indigenous peoples who have lived in place for many generations without degrading their home."[89] He cites Kiowa writer N. Scott Momaday to contrast utilitarian European approaches with Native American reverence for the land derived from a deep attunement to place. He suggests, moreover, that "informed by a sense of natural history, we should be ready to learn human history, tales that have been told, the songs that have been sung; we should be able to see more clearly how the land has invited certain kinds of work and worship and community, and how in turn our presence has transformed the land."[90] Accordingly, preserving cultural diversity is analogous to preserving genetic diversity within and among the world's myriad ecosystems. "We must help them to stay on their native ground," he declares, "help them preserve their languages and skills, for their experience can enrich our common fund of knowledge about living wisely on earth."[91]

Sanders seeks out experiences that might provide a glimpse into the relationship of indigenous peoples to their traditional homelands. Traveling through the Boundary Waters Canoe Area Wilderness of northern Minnesota, for example, he contemplates the ancient paintings on rock outcrops, the same cultural legacy that Louise Erdrich explores in *Books and Islands in Ojibwe Country*. Like other literary naturalists such as Paul Gruchow who have pondered these pictographs, Sanders contemplates their relationship to place as well as their significance to the Native peoples who made them. Midwestern pictographs, much like glyphs carved into rock across the continent, beckon to us with the mystery of their origin and antiquity. Some can readily be recognized as representations of animals that still inhabit the region. Sanders speculates about their original purpose as the icons of hunters. These same species may have held symbolic meanings or served a ritual function, such as placating the animistic spirits of game to be hunted. Moreover, representations of supernatural beings, perhaps derived from trances or dreams of the spirit world, suggest totemic or even shamanistic overtones. Perhaps most intriguing are these highly enigmatic images, which are impossible to explicate without the narratives of their creation or corresponding story cycles from oral tradition.

Sanders considers what it might mean to engage today in a profoundly personal way with such a cultural legacy. "I felt a kinship," he confides, "with those vanished artists even as I sensed in their drawings their own feelings of kinship with bear and beaver and moose."[92] He imagines painting a pictograph, with traditional red pigment and binder rendered from sturgeon, on a granite face in this place, and realizes he would wish to depict himself with his daughter now paddling there together. In fact, they participate in an indigenous tradition by making offerings of tobacco in the Ojibwe manner, as he had done there with his father many years earlier. He reflects on the historical—and ecological—significance of such a gesture: "When America was all wilderness except for occasional villages and orchards and cleared fields, a journey of any distance was a serious undertaking, whatever your ancestry.... You prepared yourself with a song and sacred tales and prayers. Traveling through forest or desert, over prairies, along rivers, you could never forget your life depended on the wild web of creatures."[93] After all, our notion of "kinship" encompasses more than personal ancestry; it includes cultural and ecological legacies precious enough to pass from one generation to next.

Sanders has also contemplated rock art in other places, such as petroglyphs along the Snake River at Buffalo Eddy near Lewiston, Idaho. Figures and designs were chipped into outcrops of volcanic basalt there as many as forty-five hundred years ago. There are "burly four-legged animals that might represent buffalo, and a scattering of slender hornless ones that might represent elk or deer. We find a zigzag design that could be a bolt of lightning, a snake, or the meandering river itself." And there are concentric circles suggesting celestial bodies such as the sun or moon, or even "the circling movement of seasons or stars."[94] But images of favored game such as bighorn sheep predominate, suggesting a connection to the animistic dimensions of such ancient hunting cultures, and a host of human forms, some adorned with horns that he speculates, cautiously, "might be men wearing masks...or shamans taking on the guise of other animals, or hybrid creatures from the time before humans and other living tribes drifted apart."[95] Yet like others who have pondered rock art such as Erdrich,

Gruchow, and Dodd, he offers such interpretations tentatively and with humility, mindful of how their origins and purpose have been obscured by the loss of mythic traditions that imbued such figures with symbolism and significance. Even establishing the age of each glyph is difficult, and archaeologists must conjecture what such "images tell about hunting practices, religious rituals, battles, or beliefs."[96] Indeed, Sanders concludes, "All we can say for certain is that the people who garnered their livelihood from this river felt the need, sustained over several thousand years, to leave a record of their actions, observations, and thoughts precisely here.... By taking in these expressions of what our ancestors saw and thought and felt, our own minds are enlarged, our sympathies expanded."[97]

In this view, traditional narratives articulate an ancient relationship to the natural world based on reciprocity and respect. As Barry Lopez argues, the perpetuation of such oral traditions reasserts a primal bond with the land. Sanders underscores the spiritual significance of such cultural memory, pointing out that "sacred stories arise from our intuition that beneath the flow of creation there is order, within change there is permanence, within time there is eternity.... The Apache word for myth means literally 'to tell the holiness.' By telling the holy, sacred stories ground people or the individual, not merely in a landscape, but in the power that creates and preserves the land."[98] Such stories reveal our profound connections to the natural world across time: "We need to be able to recite stories about hills and trees and animals, stories that root us in this place and that keep us alive...that will help us to see where we are, how others have lived here, how we ourselves should live."[99] Wary of misreading ancient images such as petroglyphs, however, Sanders reflects, "I see everything through the lens of my time, and this is a time of human-caused extinctions and disruptions.... Only in the last half century or so have we developed the capacity to extinguish, not merely a species or a tribe, but whole floras and faunas... disrupting all the life-sustaining processes on Earth."[100]

"In the world of hunters, gatherers, herders, and planters...in traditional cultures," Sanders asserts, "people who lived on the bounty of nature understood the need to protect the source."[101] He cites the example of the Iroquois, who historically deliberated on "the effects of their decisions on the descendants who will dwell in their homeland seven generations into the future," and thereby "acknowledge that membership in community entails responsibility for preserving the health of a place."[102] Contrasting Iroquois traditions of long-term stewardship with the expediency of modern commerce and consequent disregard for environmental outcomes, Sanders asks, "Looking seven generations into the future, could a developer pave another farm? Could a farmer spray another pound of poison? Could the captain of an oil tanker flush his tanks at sea?"[103]

As Sanders observed in *Writing from the Center* in the 1990s, "In America lately, we have been carrying on two parallel conversations: one about respecting human diversity, the other about preserving natural diversity. Unless we merge those two conversations, both will be futile. Our effort to honor human differences cannot succeed apart from our effort to honor... [the] bewildering variety of life on earth."[104] Moreover, "It is also dangerous to separate a concern for human diversity from a concern for natural diversity. Since Europeans arrived in North America," he reminds us, "we have been recklessly drawing on the beaver and bison, trees and topsoil, petroleum, coal, iron and copper ore, clean air and water."[105] In his view, social justice ultimately depends on protecting the ecological integrity of the natural world. He expresses this principle as a maxim in *A Conservationist Manifesto:* "Concern for ecological health and concern for social justice are therefore inseparable."[106]

For Sanders a heightened appreciation for the natural world also has compelling moral implications: "We need to enlarge our conscience so as to bear, moment to moment, a regard for the integrity and bounty of the earth. There can be no sanctuaries unless we regain a deep sense of the sacred, no refuges unless we feel a reverence for the land."[107] He extends this philosophy as a rationale for preserving those islands of natural habitat that have escaped disturbance: "In America today, the only lands with any chance of remaining wild are those we have deliberately chosen to protect. We need such lands... as we need meditation and prayer, to call us back to ourselves, to remind us of who we are and where we dwell."[108]

Yet, much like Leopold, "Sanders advocates reinhabiting a location in North America [and] learning about its natural and human history for the sake of citizenship as well as spiritual growth."[109] In a sense, Leopold's *Sand County Almanac* anticipated a central question perplexing environmentalists to the present day: how much human intervention in the form of stewardship and more aggressive management is warranted to preserve—as if in amber—representative examples of a wide variety of habitats, presumably in an idealized, "pristine" state. This approach has been likened to curating museum collections of specimens as opposed to allowing ecological processes to operate unimpeded. Writing in 1987, pioneering ecocritic Peter

Fritzell characterized this philosophy as "letting nature take its course, by leaving nature to its own devices, or by simply appreciating nature and its workings in all their deep evolutionary power and ecological complexity, wherever they may lead."[110] But what if the cumulative impact of human use seems to be leading inevitably to diminishment? To the steady decline of biodiversity as well as overall degradation of self-perpetuating ecological processes, such as when communities of native plants are threatened by invasive species? Contemporary restoration ecologists, Sanders notes, are faced with the dilemma of determining what to preserve, and how to do so, and ultimately what purposes will be served: "Once those exemplary habitats were identified, should they be artificially maintained," he asks, "in a more or less static condition, even though wild landscapes change over time?"[111] At stake is the role we should play in defending, maintaining, and restoring ecological communities compromised by disruption. "Conservationists have often been accused of wishing to freeze the land in some favored condition—for example, the American continent as it was before European colonization," he observes. "But we now realize that no such stasis is possible, even if it were desirable. If flux is the nature of nature, however, we still must make distinctions among the *kinds* of change... anthropogenic change—for example, the acceleration of extinctions due to pollution, habitat destruction, climate disruption."[112] Now that the ripples of human activity have pervaded the entire biosphere—particularly in light of the cascading ecological consequences of climate change—how are we to go about reversing dire environmental trends?

If there is a chance to dispel the prospect of environmental apocalypse, it may well be a now-or-never proposition. Accordingly, the restoration of biodiversity is elevated in our time to an ethical imperative to preserve self-sustaining ecosystems with as full a complement of native species as possible. Echoing Leopold, Sanders reminds us that "from an ecological perspective, each species is vital because it embodies an irreplaceable store of knowledge accumulated over millions of years, and it interacts with other species in ways far more intricate than we could ever fathom let alone recreate."[113] In recent essays such as "Mind in the Forest," Sanders recapitulates the continent's environmental history and poignantly laments the cumulative extent of ecological havoc: "Our indifference to other species, and even to our own long-term well-being, is demonstrated everywhere one looks, from the depleted oceans to the heating atmosphere, from poisoned wetlands to eroding farmland and forests killed by acid rain"; yet he also offers a tentative note of hope: "And yet it's equally clear that we are capable of feeling sympathy, curiosity, and even love toward other species and toward the earth.... What role might it play in moving us to behave more caringly on this beleaguered planet?"[114] If we are to go on celebrating ancient species such as sandhill cranes, for example, we must take responsibility for safeguarding habitat, including the places they breed and overwinter, as well as wetlands where they feed during their epic migration. As Sanders sees it, "These birds have flown thousands of miles... and they have been doing so, generation after generation, since the retreat of the glaciers." Like Leopold and Errington, he expresses hope that "we might be moved to treasure cranes and therefore to defend wetlands they need for survival."[115]

Perpetuating the legacy of pioneering ecologists, Sanders seeks to reimagine *stewardship:* protecting the biological processes that sustain us along with every other living species—a mission he believes must ultimately be grounded in reverence. Pondering Thoreau's dictum that "in wildness is the preservation of the world," Sanders concludes, "'Wildness' here is usually understood to be wilderness. But I think it has a larger meaning.... 'Wildness' is literally 'preservation of the World,' because without it there would *be* no world."[116] He cautions about theorizing in ways that distract us from decisive action in defense of the living earth and ultimately undermine the cause of conservation itself: "The notion of biological species is only a social construct, and so is the notion of wilderness, and so is the idea of nature itself. If our language for speaking about the wild economy of the world is suspect, then any defense of 'species' or 'wilderness' will be equally suspect, and so will the stories we tell about 'nature.'"[117]

Sanders recognizes, moreover, that our own well-being as a species—indeed, our very existence—is inextricably connected to that of the natural world, including what Leopold viewed as land health: the integrity of dynamic and self-perpetuating ecosystems. Yet, in his estimation, there is a spiritual dimension in our connection to nature as well. "Every remnant of wilderness," he declares, "is a reminder of our origin and our true home.... Through honoring both Sabbath and wilderness, we renew our contact with the mystery that precedes and surrounds and upholds our lives... a reality deeper and older and more sacred than our own."[118] Indeed, our time in such places, he reminds us, might be likened to meditation or prayer.

Sanders sees an ethical imperative for safeguarding life, human and nonhuman, "in all its dazzling array."[119] Above all, he writes, "diversity of living things... is a moral vision, one informed by an understanding of ecology and a reverence for life."[120] Enacting such a philosophy, of course, entails monumental responsibilities: "In our time we're called on to protect the soil, the air, the rivers and lakes, the forests and prairies, and the whole magnificent array of wild creatures."[121] To redress the environmentally disruptive excesses of industry and agriculture, as Sanders advocates, will necessitate thinking about our relationship, and responsibility, to nature in a fundamentally different way, one Leopold would have readily embraced. This begins with no less than "reimagining... our place in creation."[122]

NOTES

1. Sanders, *Earth Works*, 306.
2. Sanders, *Conservationist Manifesto*, 172.
3. Ibid., xii. An appendix identifies twenty seminal works of environmental literature recommended for further reading.
4. Ibid., 172.
5. Sanders, *Earth Works*, 306.
6. Sanders, *Conservationist Manifesto*, 179.
7. Ibid., 140.
8. Ibid.
9. Sanders, foreword to *Journal of a Prairie Year*, vii.
10. Sanders, *Conservationist Manifesto*, 129.
11. Sanders, *Staying Put*, 74.
12. Ibid., 75.
13. Ibid., 77.
14. Audubon lived on the Ohio River at Henderson, Kentucky, for a decade and passed through the area repeatedly in subsequent decades, including on the Missouri River expedition in the 1840s.
15. Sanders, *Conservationist Manifesto*, 130.
16. Sanders, *Staying Put*, 81.
17. Ibid., 82.
18. Sanders, *Writing from the Center*, 61.
19. Sanders, foreword to *Journal of a Prairie Year*, vii.
20. Sanders, *Conservationist Manifesto*, 179.
21. Ibid., 186.
22. Ibid., 185.
23. Ibid., 190.
24. Ibid., 211.
25. Ibid., 186.
26. Sanders, *Writing from the Center*, 44–45.
27. Ibid., 48–49.
28. Sanders, *Earth Works*, 297.
29. Sanders, *Conservationist Manifesto*, 37.
30. Ibid.
31. Ibid., 186.

32. Ibid., 37.
33. Sanders, *Writing from the Center*, 47.
34. Sanders, *Conservationist Manifesto*, 8.
35. Ibid., 129.
36. Ibid., 137.
37. Sanders, *Country of Language*, 20.
38. Ibid., 23–24.
39. Sanders, *Private History of Awe*, 26.
40. Ibid., 47.
41. Ibid., 66.
42. Sanders, *Staying Put*, 12.
43. Sanders, *Private History of Awe*, 84.
44. Sanders, *Staying Put*, 4.
45. Ibid., 12.
46. Sanders, *Private History of Awe*, 43.
47. Ibid., 69.
48. Sanders, *Country of Language*, 11–12.
49. Sanders, *Staying Put*, 8.
50. Ibid., 8.
51. Ibid., 9.
52. Ibid., 150.
53. Sanders, *Writing from the Center*, 8.
54. Sanders, *Earth Works*, 297.
55. Leopold, qtd. in Sanders, *Earth Works*, 309.
56. Backes, *Wilderness Within*, 247.
57. Sanders, *Conservationist Manifesto*, 218.
58. Sanders, *Staying Put*, 161.
59. Sanders, *Writing from the Center*, 84.
60. Sanders, *Conservationist Manifesto*, 14.
61. Leopold, foreword, viii–ix.
62. Leopold, *Sand County Almanac*, 203.
63. Sanders, *Conservationist Manifesto*, 25.
64. Barillas, *Midwestern Pastoral*, 102.
65. Leopold, *Sand County Almanac*, 203–4.
66. Backes, *Wilderness Within*, 247.
67. Finch, introduction to *Sand County Almanac*, xv.
68. Philippon, *Conserving Words*, 217.
69. Callicott, *In Defense of the Land Ethic*, 223.
70. Leopold, *Sand County Almanac*, 120.
71. Sanders, *Staying Put*, 112.
72. Sanders, *Conservationist Manifesto*, 56.
73. Ibid., 55.
74. Sanders, *Hunting for Hope*, 168.
75. Sanders, *Conservationist Manifesto*, 50.
76. Ibid., 16.
77. Ibid.

78. Sanders, *Hunting for Hope*, 37.
79. Sanders, *Conservationist Manifesto*, 150–51.
80. Ibid., 145.
81. Ibid., 153.
82. Ibid., 153–54.
83. Sanders, *Private History*, 31.
84. Sanders, *Writing from the Center*, 137.
85. Sanders, *Staying Put*, 167.
86. Ibid., 13.
87. Sanders, *Hunting for Hope*, 7.
88. Sanders, *Conservationist Manifesto*, 99.
89. Ibid., 217.
90. Sanders, *Writing from the Center*, 20.
91. Sanders, *Conservationist Manifesto*, 217.
92. Sanders, *Writing from the Center*, 129.
93. Ibid., 118.
94. Sanders, *Earth Works*, 326.
95. Ibid., 327.
96. Ibid.
97. Ibid., 327–28.
98. Sanders, *Staying Put*, 154.
99. Ibid., 169.
100. Sanders, *Earth Works*, 331.
101. Sanders, *Conservationist Manifesto*, 47.
102. Ibid., 49.
103. Sanders, *Staying Put*, 119.
104. Sanders, *Writing from the Center*, 137.
105. Ibid., 82.
106. Sanders, *Conservationist Manifesto*, 212.
107. Sanders, *Writing from the Center*, 64.
108. Sanders, *Conservationist Manifesto*, 164.
109. Barillas, *Midwestern Pastoral*, 103.
110. Fritzell, "Conflicts of Ethical Conscience," 129.
111. Sanders, *Earth Works*, 301.
112. Ibid., 340.
113. Sanders, *Conservationist Manifesto*, 14.
114. Sanders, *Earth Works*, 337.
115. Ibid., 304.
116. Sanders, *Staying Put*, 139.
117. Sanders, *Conservationist Manifesto*, 77.
118. Ibid., 166–67.
119. Sanders, *Writing from the Center*, 138.
120. Sanders, *Conservationist Manifesto*, 20.
121. Ibid., 54.
122. Sanders, *Writing from the Center*, 85.

5

Poetics of Place in "North American Sequence"

THEODORE ROETHKE (1908–1963)

Chances are, your life and the history of your place
are braided with the current of a river. . . .
Near the river you are never far from an awareness of time.

—SCOTT RUSSELL SANDERS, *Staying Put*

a course of a river that turns, moves on,
doubles back, and comes full circle,
forever arriving

—OCTAVIO PAZ, *Sunstone/Piedra de Sol*

While Theodore Roethke's groundbreaking early volumes of poetry demonstrated a striking attunement to nature instilled during his youth in Michigan, this sensibility would not reach its ultimate expression until his final collection, *The Far Field*, and particularly the ecopoetic epic "North American Sequence." Imaginatively traversing the entire continent, the sequence embodies a poetics of place, consolidating a lifetime's keen observation of nature, configured here as a transcontinental trek. In the course of this "journey" spanning the Great Plains and the Teton Range, stretching from Atlantic watersheds to Pacific headlands, these poems, replete with detailed depictions of flora and fauna, portray a remarkable array of terrains. Roethke recognized that wetlands tend to be especially rich biologically, a wellspring of abundance brimming with biodiversity. This was precisely the kind of ecological principle that pioneering wildlife biologists such as Paul Errington discerned in the marshes of the Upper Midwest. Such transitional zones, which have special appeal for both ecologists and literary naturalists, became a touchstone for Roethke's poetics of place. The expansive sequence also repeatedly addresses cultural history in relation to the continent's varied landscapes, particularly the legacy of America's indigenous peoples. Whether expressing reverence or simply celebrating what evokes wonder, Roethke found lyrical verse conducive to meditating on spiritual dimensions of our relationship to the land and its history. Titling a poem as a meditation, as he does "Meditation at Oyster River," reminds us of his intent to depict moments of intuitive communion with aspects of the environment that might still be deemed wild.

"North American Sequence" was central to establishing Roethke's reputation among major American poets at midcentury. He clearly aspired to make a lasting contribution and a bid for a place in literary history, venturing into ambitious forms for the sequence: six poems, each composed of three to five sections, together forming a compact epic of self-discovery and direct experience of the sublime in relation to nature. By the time his final collection was published posthumously in 1964, Roethke was already a celebrated figure, having been awarded the illustrious "triple crown" of poetry prizes: the Pulitzer, National Book Award, and Bollingen. He received a second National Book Award for *The Far Field.* Still "North American Sequence" was ultimately hailed by critics as his masterpiece. In fact, the inclusiveness of "North American Sequence" has led many critics (e.g., Balakian, Barillas, Mills, Parini) to compare it favorably to Walt Whitman's *Leaves of Grass.* Critic Ralph J. Mills Jr., for instance, declared Roethke's sequence his magnum opus, containing "a number of the most astonishing mystical poems in the [English] language."[1]

Such accolades—as well as critical reception over time—are inevitably subject to shifting literary fashions and taste. Roethke was among poets of his generation at the forefront of a transformation in American verse, as formal tone was supplanted by the ascent of a more colloquial idiom exemplified by William Carlos Williams; as personal experience was deemed legitimate subject matter by Confessional poets such as Robert Lowell; and as organic free verse eclipsed patterned forms of the past. In this regard, ecocritic Bernard Quechenbach points to Roethke's second book, *The Lost Son and Other Poems*, as pivotal, effectively ushering in the contemporary period in American poetry and, moreover, deems Roethke the most significant *nature* poet of his generation.[2]

Yet from an ecocritical perspective, another profound cultural shift had long been under way. Our once largely agrarian society has grown progressively more urban and suburban, as argued so compellingly by Wendell Berry in *The Unsettling of America*. Consequently, over the course of a century or more, personal contact with nature has sharply decreased, as even the *memory* of subsistence family farming has faded. Bill McKibben warns in *The End of Nature* that ultimately such a "loss of memory will be the loss of meaning."[3] Similarly, the esteemed elder statesman of what has recently come to be known as ecopoetry, W. S. Merwin, laments that many contemporary readers are no longer equipped to fully fathom the symbolism of images corresponding to the natural world—let alone recognize their emotional resonance—on which Roethke's poetics of place squarely rests. Similarly, Quechenbach asserts that "nature writing, and by extension nature poetry, serve as a kind of record of the culture's attempts to comprehend and coexist with nature," the textual equivalent of ecological memory.[4] During the first decades of the twenty-first century, we read Roethke at a time when environmental decline has been greatly exacerbated by climate change, perhaps irreversibly so, by undermining the integrity of self-regulating ecological processes essential to entire biomes. We need more than ever to recognize what must be seen as sacrosanct, inviolable, and fundamentally a part of us. This vision underlies the enduring significance of Roethke's work.

Roethke was at the pinnacle of his creative powers when composing the poems of this sequence, a tribute to the continent's monumental geography, diverse wildlife, and indigenous history—positioning himself as "an aboriginal trailblazer, making a path through the raw wilderness."[5] The sequence first appeared in his final collection, whose title, *The Far Field*, harkens back to agrarian landscapes of the Midwest, places imbued with personal history. Realizing that it was likely to be his final book, he reportedly worked feverishly on the manuscript during the very last days of his life.[6] He confided as much in poems such as "Infirmity," and even more emphatically in "The Far Field" itself. Contemplating mortality undoubtedly lent the project personal urgency and inspired the lofty ambition of composing a single sequence capacious enough to tie together a lifetime's worth of meditations on the natural world.

While this audacious work spans the entire continent, a countervailing force of memory leads Roethke to return repeatedly to places from his youth. Place-names echo throughout the sequence, tracing a trajectory across the American West and grounding passages in specific locales, but also invoking landscapes from the Midwest in the process. As Indiana essayist Scott Russell Sanders has observed, the landscapes of childhood form a fundamental frame of reference for our later appreciation of the natural world in other places.[7] In this sense, Michigan forms the bedrock of Roethke's perception of environmental history. And while he eventually taught at the University of Washington, his roots as a *midwestern* writer undeniably remained deep.[8]

Even as "North American Sequence" progresses westward, suggesting a transcontinental journey, memories of the Midwest well up forcefully; places that had left a lifelong impression.[9] Those landscapes internalized as a child continued to provide grist for metaphysical musing until the end of his life. As Peter Balakian explains: "memory becomes mythopoetic in its ability to bring past and present into a moment of convergence."[10] The sequence as a whole juxtaposes many vistas that ultimately underlie his perception of place, what ecocritic William Barillas characterizes as "a sweeping view of the continent, from the Pacific Northwest to the Great Plains and the poet's beloved Michigan... and for mystical union with the forms of nature—plants, animals, rock, and water—that he had encountered in particular places in North America."[11] As critic Jenijoy La Belle suggests, "These east and west journeys recall the physical movement... but... the cardinal concern is with a movement in the other direction... to where [he] came from—the Midwest. These journeys into the interior—both of the self and of the continent—have a temporal as well as a spatial dimension."[12] Above all, the trope of the journey came to represent an exploration of the self in relation to nature on a metaphysical level. Indeed, "Roethke in a very important sense *is* the North American continent," Malkoff claims, "and his explorations of it in time and space continue his search for the self by retrieving the past and analyzing the present."[13] Melding a medley of places gleaned from a lifetime of memories, Roethke's method in the sequence might even be likened to how Thoreau combined several years of experience at Walden into a single, seasonal round.

Roethke detailed his midwestern beginnings and personal affinity for the region in his essay "An American Poet Introduces Himself and His Poems": "Everyone knows that America is a continent, but few Europeans realize the various and diverse parts of this land. The Saginaw Valley, where I was born, had been great lumbering country in the 1880s. It is very fertile flat country in Michigan.... It was this region that my grandfather came to in 1870 from Prussia, where he had been Bismarck's head forester."[14] Introducing himself as a writer first and foremost in terms of geography and history, his interpretation of North American landscapes is situated in the colonial narrative of settlement in the Upper Midwest. Europeans like his grandfather were drawn by the abundance of natural resources such as seemingly inexhaustible tracts of "virgin," old-growth timber. Within two generations, the face of this landscape changed irrevocably as primeval stands fell and verdant wild ecosystems were replaced by second-growth forest or cleared for cropland, at a terrible ecological cost: "the loss of ancient woods, the destruction of rivers, the extinction of animals."[15] By the time of Roethke's birth in 1908, that boom had run its course: "By 1900, it was all over. A few virgin stands of timber were left, islands in a sea of stumps."[16] Accordingly, his account of landscape change encompassed both old-growth forests and subsequent second-growth woodlands returning in the wake of logging. Such juxtapositions reveal environmental history as indigenous trees regenerating naturally demonstrate ecological memory in its primary, biological sense. In fact, a host of native tree species would emerge as a leitmotif in his poems, appearing nearly twenty times in "North American Sequence" alone.[17]

Allan Seager opens his biography of Roethke, *The Glass House*, with a lyrical evocation of the poet's birthplace taken from Alexis de Tocqueville's description of travels on the Saginaw River in 1831 in *Journey to America*. De Toqueville describes the region's seemingly primordial forests in reverent tones (reflecting the dogma of his day): "The river watered an immense forest.... The wilderness was before us, just as six thousand years ago it showed itself to the father of mankind."[18] In Seager's estimation, Roethke's sensibilities developed in relation to this landscape and his family's history there, as well as its storied past:

> He knew the "native" in American, the untamed in the dark wood, the primitive in the Indian, for he was of their world, as his Old World fathers tended virgin timber (the last pure Michigan stand in the First World War) and replanted forests felled by pioneers.... The poet worshipped in the New World, a native, at the axed roots of virginal trees cleared for pastures. And the trees dwarfed him as they had staggered Columbus.[19]

Having gained an appreciation for the environmental history of the Midwest while growing up in Michigan, he understood how development could transform entire ecosystems. As William Barillas points out, "Roethke was well aware of how drastically his home region had been altered in the half century preceding his birth."[20] He described his family's own holdings as including twenty-five acres "in the town, mostly under glass and intensely cultivated," plus a remnant of primeval forest, "the last stand of virgin timber in the Saginaw Valley" to have survived Michigan's timber boom, and "a wild area of cut-over second-growth timber, which [his] father and uncle made into a small game preserve."[21] In the poem "Moss Gathering" from his second collection *The Lost Son and Other Poems* (1948), he describes tearing up moss as if it violates an ancient order, showing an awakening sense of environmental conscience. Having descended from among the earliest and most primitive land plants to have evolved more than four hundred million years ago, the moss he gathered was now destined to line cemetery baskets. The process of landscape change encapsulated here would appear repeatedly in the crowning achievement of his distinguished career, "North American Sequence," an expansive poem expressing a transcendental appreciation for wilder places and reflecting an awareness of forces that had shaped them.

In composing "North American Sequence," Roethke also consciously responded to *literary* history in a number of ways. There are Whitmanesque long lines and catalogues, for instance, as well as philosophical resemblances.[22] Roethke himself readily conceded such debts by addressing Whitman directly in "The Abyss." Echoing the tenets of Romanticism, moreover, he denounced reckless and excessive industrialization and expressed revulsion at the desecration of nature by deforestation. In fact, the sequence opens with an evocation of urban industrial blight, reflecting a neo-Romanticism shared with other twentieth-century American writers from the Midwest, including his own student James Wright. Harkening back to the approach of British Romantic poets such as William Blake and William Wordsworth, he sought direct experience of the sublime in nature: expressing at once awe for nature's majesty and terror before its power. Roethke himself acknowledged such affinities. Indeed, at moments there are discernible echoes of Blake's work.

The sequence repeatedly shifts time frames, frequently evoking *remembered* places and landscapes. Roethke juxtaposes past and present in a manner that resembles "Lines Composed a Few Miles Above Tintern Abbey," in which Wordsworth reconceives place through a layering of time. Though "Tintern Abbey" was written in 1798, Wordsworth anticipates what in the following century would become known as Transcendentalism, which sees nature as a quasi-divine source of guidance akin to scripture, intimating a kind of animism. A sense of timelessness exists in the most mystical experiences conveyed in the sequence, in poems such as "Journey to the Interior."

By invoking primitive life forms such as lichen and lizards, Roethke suggests an ancient stasis, an eternal moment. Similarly, he recognized a timelessness in the return of migrating songbirds, such as the arrival of warblers at the beginning of May. Such images convey sentiments traditionally associated with Romanticism, elevated by sheer lyricism, while their immediacy suggests a transcendental sensibility. Throughout the sequence, Roethke's contemplative meditations on nature echo the metaphysical inclinations of such American Transcendentalists as Ralph Waldo Emerson and Henry David Thoreau. In fact, Roethke acknowledged having carefully read and contemplated the Transcendentalists (as well as Whitman), and reported having purchased personal copies of both Emerson and Thoreau while still an adolescent.[23]

Roethke's connection to nature achieves an intimacy when he recounts visceral encounters with place—often landscapes rendered with the particularity of lived experience refracted by memory. Subtle images are composed with stunning precision that Jay Parini construes in transcendental terms: "clarity of vision . . . heightening of perception generally: a sense of true significance of the physical world [as] inner and outer realms come into Emersonian alignment."[24] In fact, metaphysical ruminations throughout the sequence are often grounded in corporeal relationship to topography in ways found in Loren Eiseley's writing, like witnessing the primal force unleashed when river ice gives way all at once to signal the end of winter.

"Meditation at Oyster River," for instance, pivots deftly between a delicate depiction of meltwater in a little creek and the unforgettable ferocity that mesmerized the poet as he witnessed the breakup of river ice in spring—its awe-inspiring, primal power. What elevates such detailed catalogs above description, beyond being so keenly observed, is how the riparian terrain becomes a topographical analog for the poet's inner searching, much the same Transcendental territory A. R. Ammons would explore so compellingly a generation later. Above all, Roethke found *rivers* conducive to viewing through the prism of his lyric verse. Their expansive geographical sweep seems to evoke a whole watershed, perhaps an entire biome. Moreover, depictions of place are correlated with the motion of the poet's mind perceiving it, a landscape characterized by confluences and currents, at once aqueous and transitional.

Such exquisitely detailed accounts might be likened to those of contemporary literary naturalists working in prose. The difference is that Roethke was constantly seeking deeper philosophical realizations, much like Thoreau, and ways of tracing the motion of his own mind. In his critical study of *The Far Field*, poet Peter Balakian characterizes the poet in just such terms: "a visionary seeker who must also be a cataloging naturalist.... A Thoreauvian naturalist, Roethke creates the rhythmic beauty of North America's indigenous creatures."[25] Balakian portrays Roethke as engaging in what he terms the "creation of the self as representative man and visionary explorer, and in his conception of America as a primal territory that is at once a realm of consciousness and geographic space."[26] Indeed, his way of contemplating metaphysical dimensions of engaging with nature often resembles that of a mystic, grounded in the particularity of place.

Roethke's evocations of landscapes often take an inward turn. A series of fragments from his copious notebooks published posthumously in *Straw for the Fire* provide a sort of ars poetica revealing how landscape inspired his writing: "The first snow transformed the whole field. The ploughed loam, with its high ridges left in place. One bird at the center of the tree.... That sense: here I belong—so rare.... Seminal states of waiting, a sacramental view of nature, listening to wind-sounds, looking at snow in the far-away field. A sense that from each image some profound truth might be grasped."[27] We hear a deep-seated yearning to engage place at an elemental level by encountering the natural world in a primal way.

Yet his way of seeing and conceiving nature was also informed by environmental history. Later in the sequence, as the topography of fields and cliffs gives way to the aqueous images of tidal wetlands, lakes, and rivers, he emerges as a literary naturalist attuned to the dramatic processes shaping the landscape. He was particularly drawn to places such as confluences and estuaries where merging currents and blurred boundaries provided apt metaphors for his metaphysical yearning and transcendental appreciation for the "spiritual significance of nature."[28] While entirely devoid of church doctrine or denominational dogma, "North American Sequence" might be likened to a personal pilgrimage consonant with the expression of mystical experience in many religious traditions. As James Dougherty observes, "the Sequence traces a spiritual process that is founded on his recognition of the substantive presence of land and sea, the birds and plants."[29] He gravitated toward the most dynamic landscapes, often expressing a sense of the ineffable, a "place of change and exchange, a territory...where life undergoes perpetual transmutation."[30] And the intricacy of such images as tide rippling into an estuary is complemented and intensified by the eloquence of his language.

Biodiversity beckoned to him from such wetlands. The sequence repeatedly recites a litany of life forms on the coastline, from tiny fish and crabs on the beach to gulls, ravens, deer, snakes, sandpipers, hummingbirds, and kingfishers along the shore. Many of these catalogs are so closely observed that they might be likened to the meticulous descriptions in field guides intended to help one identify species. Intertidal wetlands, where riparian habitat intersects with the seashore, are noted for exceptionally dynamic ecosystems, what Kalaidjian calls "a landscape of ecological transition stages."[31] Similarly, literary naturalist Barbara Hurd celebrates what she terms the *ambiguity* of such landscapes. "No places are more ambiguous than swamps and bogs. Their identity wanders between solid and liquid, sloshes back and forth over the line between firm and yielding.... But a liquid landscape cannot be nailed down with maps and charts.... What's between wet and dry land is a broader border, a wavy intermediate zone whose boundaries don't hold still.... When, by some act of grace, the lines we think are there dissolve, something else appears, something timeless and rich, an intermediate zone, languid and latent, the lushness of something about to be." In a word: fecundity.[32] Other poems in *The Far Field* such as "All Morning" revel in resulting species biodiversity, naming more than a dozen species. "Forming a montage of memories suspended in time, seasons are transmuted into the patina of eternity." Yet Roethke recognized that seemingly timeless abundance is always shadowed by the risk of extinction, such as suffered by the Carolina parakeet and passenger pigeon. Given the course of environmental history, Roethke's declaration reminds us to savor places that retain a semblance of wildness, especially those that provide ample habitat for wildlife that once flourished there.

In addition to naming an array of indigenous flora and fauna, the poems of the sequence also portray complex organic webs, whether a pine tree sinking into an estuary or an osprey perched in its empty branches surveying a river mouth teeming with trout and salmon. Even processes of decay seem palpable: a wave ripples toward a shattered tree, shells lie scattered across the sand. Recurrent images of water add cohesion to the sequence. At moments Roethke identifies with the *motion* of water as a natural force shaping and reshaping the landscape, such as immense stones streaked by the high water of previous seasons. He often renders a watercourse with such intricacy as to suggest the arc of perception itself, as when he admires the agility of delicate birds alert to every threat.

In contrast to a more abstract or archetypal use of natural imagery in Roethke's earlier work, "North American Sequence" portrays ecological niches with great particularity. Certain recurring images such as birds still take on symbolic overtones, but are connected to the ecological dynamics of the places they inhabit. A stationary raven surveying the river from its roost contrasts with the blur of a hummingbird among blossoms. Consider how the final poem of the sequence, "The Rose," opens with a lyrical evocation of avian abundance at the edge of the Pacific Ocean: heron and towhee, finch and kingfisher, a line of geese and the solitary owl. Several celestial cycles seem to intersect in this moment: from the diurnal transit of the sun to the lunar tug of tide. Again and again, the poet makes a mystical leap, identifying spiritually with nature and place and particularly with birds that so often seem animated with sentience and purpose.

Identifying such a wide variety of plants and animals demonstrated Roethke's increasing knowledge and interest in ecology; similarly "using the westernized vocabulary of arroyos, buttes, and canyons, the poet conveys an immediate, particular encounter with the land."[33] Yet he also recognized the dynamic process of ongoing landscape change under way, such as geographical forces that perpetually cause erosion, where dry creek beds and arroyos suddenly rage with floodwaters following torrential rains, chiseling at the base of buttes, carving the canyons. The sequence is expansive in its geographic reach, extending from where Oyster River empties into the Atlantic to the headlands of the Pacific Northwest, constituting what Kalaidjian calls a "consummate celebration of American place."[34] The poem's disparate and far-flung landscapes include the Upper Midwest, the Great Plains, and the Teton Mountains.

Roethke's initial acquaintance with the geography of the West came when he crossed the grasslands, Great Plains, and western mountain ranges by train. The intensity of his encounter with these landscapes was undoubtedly heightened when he drove across them in 1950. The trope of a journey west that lends structure to the sequence also symbolizes the poet's personal quest across such a storied landscape. A transcontinental journey westward inevitably recalls the most iconic of American narratives—wave after wave of explorers and soldiers, trappers and mountain men, pioneers and gold miners headed west in the wake of Manifest Destiny. The slogan "Go West" still evokes such folk history, and with it a mythology of discovery and renewal. "North American Sequence" is clearly heir to this national narrative, and with it the counter-narrative from a Native American perspective.

There are a number of reasons to believe that Roethke envisioned the project in this way himself. "As early as 1944," Rosemary Sullivan recounts, "he was writing to William Carlos Williams of his need for a device to organize his ideas about the American Midwest. His sequence was to be a celebration of the landscape."[35] Further evidence can be found among his published letters, including a proposal in 1959 for funding from the Ford Foundation that suggests how he had begun to conceptualize what was to become in part "North American Sequence." What he proposed was essentially a revisionist history: "A sequence of serious poems beginning with a long dirge which will express through suggestive and highly charged symbolic language the guilts we Americans feel as a people for our mistakes and misdeeds in history.... This would not be chronological, yet would expose some of the lies of history."[36] Harry Williams argues that the historical "tragedy of America" that Roethke meant to revisit necessarily includes the Indian Wars of the nineteenth century such as the massacre of the Oglala Sioux at Wounded Knee, South Dakota. Williams concludes that "there is a sense of guilt in the poem for a broken nation, and Roethke seems to have had a collective sense of guilt in mind in composing 'North American Sequence.'"[37] Immersed in reading American history, Roethke contemplated a second major poetic sequence predicated on a journey to visit the sites of Indian betrayals and defeats, which he clearly viewed as a moral travesty and a national disgrace. As Seager explains, late in life "he had been reading history about the Indians, for there are references [in some of his last notes] to Custer, Crazy Horse, General Crook, Black Elk, and Chief Joseph. If we think in terms of the growth of Ted's mind, it seems to have taken nearly his whole lifetime to come to terms emotionally and spiritually with the presences of the Saginaw Valley."[38]

The watershed where he grew up is named for "a version of an Algonquin word that means 'the place of the Sauks.'"[39] Michigan's Saginaw Valley had long been Algonquin/Sauk territory, until an alliance of Ojibwe, Huron, Potawatomi, and Menominee displaced them centuries before white settlement: "Eventually all that was left of the Indians in the valley was the sibilance of their place names, Saginaw, Shiwassee, Tittabawassee."[40] These are places that figure prominently among those named in "North American Sequence." As a boy Roethke "must have been aware of the Indians," as Seager points out, "for he collected a shoebox full of flint arrowheads in his rambles along the riverbanks," although the interest in pre-contact history does not appear in his notebooks or poems until late in his life.[41] In the end, "North American Sequence" became a vehicle for exploring his own relationship to place against this historical backdrop.

Roethke sought to write a poem indigenous to America—a nation-state and its landscape, the continent and its history. Beyond rendering the continent's distinctive topography, he also evokes environmental history by summoning the legacy of the buffalo so pungently as to seem visceral: their carcasses and dung reeking under the summer sun. Given its near extinction and subsequent resurgence, the species remains for us today—as much as for the poet—the embodiment of America's environmental legacy, perhaps its very epitome. And he repeatedly invokes the figure of the Indian in wilderness settings, which has long been associated with ecological integrity in both visual and literary representations. These are the traces—and absences—that led him deeper into an encounter with the continent's majestic geography as well as its vexing past.

As Dougherty observes, "Casting back not in personal memory but in the memory of the continent, he smells the dead buffalo—possibly an Oglala hunt, but perhaps the mass slaughter of the herds that accompanied the suppression of the Sioux and other Plains peoples."[42] In fact, the opening section of "North American Sequence" culminates by naming just such tribes—Oglala and Iroquois—signaling a link to North America's indigenous past, and suggesting how our relationship to landscape, place, and environmental history continues to shape national culture. In his essay "On 'Identity,'" Roethke contends that the crux of any quest for identity comes down to a single question: "What to do with our ancestors…both the literal or blood, and spiritual ancestors."[43] His evocation of the indigenous peoples in "North America Sequence" provides at once a cultural, historical, and ecological frame of reference.

Roethke's nascent environmental sensibilities have also been likened to those of pioneering ecologist Aldo Leopold, leading critics such as Barillas to suggest that "North American Sequence" represents an extension of the land ethic originally posited by Leopold, and so demonstrates a heightened sense of "ethical and aesthetic perspective on American cultural history and geography."[44] The two clearly had similar perspectives on the region's environmental decline: "the prairie in relation to ecological history...the diminishment of natural diversity regretted by Leopold and Roethke and symbolized by both in the tragic near extinction of the American bison."[45] They recognized that the plight of buffalo epitomized the ecological havoc caused by westward expansion across the continent during the nineteenth century. Such themes of "environmental destruction and genocidal assault on native peoples" remain central to his legacy.[46]

Equally aware of his literary heritage, Roethke echoes T.S. Eliot's dictum from "The Wasteland" that "Old men should be explorers" by likening himself to an Indian at the close of "The Longing."[47] One reading of this rejoinder holds that Roethke concurred with Eliot; that is, the Indian is taken to be a type of explorer and is likened to a "pioneer," a "backwoodsman," a "native explorer," and an "aboriginal."[48] Yet, as Barillas concludes, "By refusing to adopt Eliot's explorer archetype, he rejects the expatriate poet's Anglo-Catholicism; by assuming that of an Indian, he professes an autochthonous[49] spirituality of place."[50] In fact, the poem can be read as *revising* Eliot. The closing lines of the sequence signal a profound contrast between the two poets: Eliot the expatriate living in Europe versus Roethke's identification with the landscapes of North America. As Barillas contends, "Eliot abstracts place, observing North America at a distance rather than involving himself intimately with the continent's history and landscape," whereas "at times [Roethke's] is the eye of the naturalist."[51] There is also a profound difference in perspective between Eliot's devotion to the dictates of the Anglican Church and Roethke's animistic regard for nature as "a native namer, reclaimer of territory, the perpetual explorer"[52] for whom "sacredness has nothing to do with theology, and everything to do with biology, evolution, and the primeval life force."[53] The poet characterized his own attitude as primitive and animistic: "In a quest for identity," he summoned "*all* living things." Moreover, "everything that lives is holy," he declared. "I call upon these holy forms of life."[54]

Throughout his career, Roethke's intimations of communion with nature reveal a transcendental bent when contemplating creatures as dissimilar as hummingbirds, deer, and snakes. Kalaidjian characterizes this stance as transcendental: "nature speaks to the poet through its indigenous plants and animals."[55] A timeless mystical union is articulated in his poem "Snake," for instance. At such moments, it is as if the observed and observer become one, what has been aptly termed *sacramental imagination*. The behavior of bees, worms, minnows, and butterflies—life forms regarded as simple creatures in the classical Great Chain of Being—remain for the poet deeply mysterious, possessing sensory powers that may exceed our own and beyond the reach of human reckoning. Even a lowly slug in his garden triggered empathy and stirred his imagination, suggesting that weasels and bats or even toads and worms were somehow kindred to him. Roethke's student poet David Wagoner described it as "an immediate, seemingly effortless metamorphosis into the relatively benevolent subhuman."[56] As the poet himself explains, "the identity of some other being... brings a corresponding heightening of one's own self, and, even more mysteriously, in some instances, a feeling of oneness with the universe... induced... by intensity in the seeing. To look at a thing so long that you are a part of it and it is a part of you."[57] Such sustained and intimate attention gives rise again and again to a kind of identification. Such shape-shifting resembles transformations between animal and human forms common to the oral story traditions of many Native American peoples, though Roethke himself does not claim that association. Rather, it is a vision born of direct encounter with place, akin to heightened moments found in the writing of Loren Eiseley, as if one is evoking memory from deep time, or perhaps summoning a reptilian sensibility. To imagine being one with each of these creatures—to the degree allowed by language, imagination, and conceivably even ontogeny as a form of corporeal memory—became for Roethke a way of perceiving and expressing his profound affinity for the natural world.

In the final poem of the sequence, Roethke again contemplates wetlands, a confluence characterized by remarkable abundance and biodiversity, where he observes a multitude of birds as if in a single glance—hawks, eagles, seagulls, herons, towhees, kingfishers, scoters, geese, and whooping cranes. The sequence as a whole names an impressive fifty-eight species and subspecies of birds alone.[58] On the Pacific headlands, having completed his transcontinental journey through memory and imagination, he is struck by the resilience of a delicate flower, the wild rose flanked by driftwood, black sand, and surf. The rose offers a locus of stability amid constant motion. Dougherty casts Roethke's depiction of the place in mystical yet decidedly ecological terms: "the rose itself is in a struggle, a state of tension with everything around it—its immediate botanical competitors, its seashore ecology, the sea-wind, the sea pine, the coastal madrona, the interpush of creek current and Pacific wave."[59] In a landscape of utter flux, he found a way to portray his own spiritual quest to embrace the natural world in all its grandeur and particularity—once more with an immediacy reminiscent of Thoreau.[60]

Roethke's vision of environmental history and a spiritual relationship to the natural world forged in Michigan reached its consummation in "North American Sequence." "Steeped as a Midwestern child in damp drooping roots," as John Felstiner concludes, "Roethke in the Pacific Northwest discovers a wild grace."[61] Such geographical meandering also introduces a complex temporal element, as he deftly maneuvers between immediate impressions of place and reminiscence that lends a profound emotional resonance to the present. In these poems he seems to have arrived at another level of self-knowing, made possible by a lifetime of journeys and a keen appreciation for place. His sense of self is not separate from the natural world but inextricably entwined with it. There is a sense of exaltation in the poet's spiritual renewal through union with nature. Roethke's place in all of creation—illuminated by an awareness of transience and his own mortality—sharpened his appreciation for the environmental history of myriad landscapes across the continent, beginning with Michigan. In the end, each had become a part of him.

NOTES

1. Stein, *Theodore Roethke*, 135.
2. Quechenbach, *Beyond the Far Field*, 31, 43.
3. McKibben, *End of Nature*, 213.
4. Quechenbach, *Beyond the Far Field*, 20.
5. Balakian, *Theodore Roethke's Far Fields*, 139.
6. For example: Parini, *Theodore Roethke*, 160–67; Bowers, *Theodore Roethke*, 148.
7. Sanders, *Staying Put*, 4.
8. William Barillas suggests that "Roethke's ties to the Pacific Coast have been overstated, his midwestern character too readily slighted" (*Midwestern Pastoral*, 136). Recollections of childhood experiences in the family's greenhouses, for example, grace the pages of his second collection, *The Lost Son and Other Poems* (1948).
9. See Balakian, *Theodore Roethke's Far Fields;* Barillas, *Midwestern Pastoral;* Sullivan, *Theodore Roethke.*
10. Balakian, *Theodore Roethke's Far Fields*, 131.
11. Barillas, *Midwestern Pastoral*, 130.
12. La Belle, *Echoing Wood of Theodore Roethke*, 155–56.
13. Malkoff, *Theodore Roethke*, 176.
14. Roethke, "American Poet," 7.
15. Barillas, *Midwestern Pastoral*, 122.
16. Seager, *Glass House*, 6.
17. Malkoff, *Theodore Roethke.*
18. De Tocqueville, qtd. in Seager, *Glass House*, 1.

19. Lincoln, *Sing with the Heart of a Bear*, 220–23.
20. Barillas, *Midwestern Pastoral*, 110.
21. Roethke, "American Poet," 8.
22. Roethke, *Selected Letters*, 230.
23. Roethke, "Theodore Roethke," 15–16.
24. Parini, *Theodore Roethke*, 163–64.
25. Balakian, *Theodore Roethke's Far Fields*, 148.
26. Ibid., 131–33.
27. Roethke, *Straw for the Fire*, 143.
28. Barillas, *Midwestern Pastoral*, 111.
29. Dougherty, "Theodore Roethke's 'North American Sequence,'" 184.
30. Balakian, *Theodore Roethke's Far Fields*, 137.
31. Kalaidjian, *Understanding Theodore Roethke*, 146.
32. Hurd, *Stirring the Mud*, 67, 69, 78.
33. Dougherty, "Theodore Roethke's 'North American Sequence,'" 189.
34. Kalaidjian, *Understanding Theodore Roethke*, 122.
35. Sullivan, *Theodore Roethke*, 167–68.
36. Williams, "*Edge Is What I Have*," 220.
37. Ibid., 114.
38. Seager, *Glass House*, 279.
39. Ibid., 2.
40. Ibid.
41. Ibid., 8.
42. Dougherty, "Theodore Roethke's 'North American Sequence,'" 188.
43. Roethke, "On 'Identity,'" 23.
44. Barillas, *Midwestern Pastoral*, 132.
45. Ibid.
46. Ibid., 138.
47. Overt allusions to and deliberate echoes of Eliot in "North American Sequence" include the use of the classical elements of air, water, fire, and earth as well as exalting in the image of the rose.
48. Balakian, *Theodore Roethke's Far Fields*, 133–35.
49. *Autochthonous*: "risen or sprung from the ground or from the soil he inhabits; one of the original inhabitants or aborigines, a native" (*Webster's New International Dictionary of the English Language*, 2nd ed.).
50. Barillas, *Midwestern Pastoral*, 133.
51. Ibid., 133, 107.
52. Balakian, *Theodore Roethke's Far Fields*, 131.
53. Barillas, *Midwestern Pastoral*, 117.
54. Roethke, "On 'Identity,'" 24.
55. Kalaidjian, *Understanding Theodore Roethke*, 139.
56. Wagoner, introduction to *Straw for the Fire*, 5.
57. Roethke, *"On 'Identity,'"* 25.
58. Kearful, "Regions of the Self."
59. Dougherty, "Theodore Roethke's 'North American Sequence,'" 194–95.
60. Ibid., 194.
61. Felstiner, *Can Poetry Save the Earth?*, 222.

6

Landscapes of the Past

William Stafford (1914–1993)

There is a source for this deep restlessness;
and the path that leads there is not a path to a strange place,
but the path home.

—Peter Matthiessen, *The Snow Leopard*

William Stafford's poetry is distinguished by a reverence for nature instilled during his upbringing on the grasslands of Kansas. His first two collections contain some of his most celebrated and enduring work: *West of Your City*, followed by *Traveling Through the Dark*, which garnered a National Book Award and established his reputation as a major voice among twentieth-century American poets. Both books were profoundly grounded in landscapes of the past and show a pronounced influence of formative experiences of place. Later writing reveals his trajectory westward. Journeys are woven into many poems, both personal and historical ones, reflecting his interest in cultural and environmental history. *The Way It Is*, for instance, a compilation gleaned from the dozens of volumes published during his lifetime and published posthumously in 1998, spans an impressive geographic range throughout the American West, including the Plains states of the Dakotas; Rocky Mountain states such as Colorado, Wyoming, and Montana; and West Coast states from California to Alaska. Yet places that had been internalized during childhood continued to inform how he perceived landscapes throughout North America. Even late in his career, recollections of the Midwest consistently well up in his work. In fact, the plains of Kansas remained a fundamental frame of reference throughout his writing life, much the way Roethke's Michigan did in "North American Sequence."

Stafford's voice as a poet is first and foremost an evocation of place, invoking both environmental and cultural history. In aggregate, his writing demonstrates a keen appreciation for the continent's varied topography; he often historicizes his accounts by incorporating references to pioneers who in settling there displaced indigenous peoples that had preceded them by millennia. In fact, *West of Your City*, his first volume of verse, opens with an unmistakable emphasis on the prairie biome of Kansas grasslands. "One Home," for example, situates his own family's oral history in the broader sweep of initial white settlement, triggering memories of Indians who long inhabited the region before being pushed westward:

> Mine was a Midwest home…
>
> outside, the buffalo grass, and the wind at night.
>
> A wildcat sprang at Grandpa on the Fourth of July
> when he was cutting plum bushes for fuel,
> before Indians pulled the West over the edge of the sky.[1]

Images of the region's native flora arise throughout these poems, such as stands of cottonwoods and wild plum bushes punctuating vast tracts of buffalo grass in a prairie biome animated by a wide array of wildlife. Indeed the litany of species encountered in the new and selected poems, *The Way It Is*, reads like a field guide, encompassing mammals such as antelope, badger, buffalo, cougar, coyote, deer, horse, muskrat, and wildcat, as well as birds such as cranes, killdeer, hawks, herons, and quail, and finally even fish, above all salmon. "Wild things wait," he recollects in "Midwest," "crouched in those valleys / west of your city outside your lives."[2] The grasslands encompassed a region where agricultural acreage intersected with vast tracts of open space. "Where I walk is road," he wrote, "but where I look around is wilderness, if I look far enough."[3] Though not the relatively extensive ecosystems of remote wilderness preserves, perhaps, natural abundance could still be readily glimpsed nearby: "beyond the Cimarron crossing / and after the row-crop land," he wrote, "a lake would surprise the country / and sag with a million birds."[4]

The first compelling evidence of Stafford's lifelong affinity for the grasslands of Kansas is provided by "White Pigeons," evidently his first attempt to compose a poem while still an undergraduate at Kansas University. This poem appears in *Another World Instead*, a collection selected from archives of some four hundred previously unpublished works dating from as early as 1937.[5]

> This is the aching land,
> The bleak and desolate.
> This is the plains.
> On this blank loneliness in huddled clump
> A house, a barn, and fences.
> A boy foreshortened, small, wind-buffeted,
> His pigeons watched come home.
> Hard sky, hard earth.[6]

This, his very first utterance in verse, immediately establishes a geographical consciousness while expressing a desire, much like Roethke's, to situate his sense of self in relation to place. In fact, introducing *The Way It Is*, Naomi Shihab Nye recognizes Stafford's penchant for contemplating place-based cultural history: "William Stafford, originally from the land of Kansas, to the land of Oregon, to the widest lands of being, was a champion of language, a seeker, a deep rememberer."[7]

Stafford's landscape narratives often evoke locations in a visceral way. "By the Snake River," for instance, offers richly textured depictions of topography as the river churns tumultuously down canyons between volcanic crags. Yet, like Roethke's "North American Sequence," such poems also tend to turn inward, exploring the process of perception itself while observing and interacting with the natural world:

> to find
> what I am, among sights given by chance
> while I seek among rocks, while the deep sturgeon move
> in this water I lift pouring through my hands.[8]

Subsequent poems in the second volume echo the place-based refrain while introducing an element of transience as well. "Prairie Town," for example, describes the ground beneath an abandoned settlement, latticed by salt mines while a river flowed beneath its now empty streets. We glimpse in the span of a single stanza the arrival and departure of pioneers on the short-grass prairie whose farms had failed within generations for lack of rain, culminating with the exodus of the Dust Bowl.[9] By moving west himself, Stafford consciously followed in the tracks of that diaspora. Many pioneer homesteads that once briefly flourished in the vast expanse of grass on the Great Plains were ultimately abandoned, such as the one Stafford describes in a poem subtitled "A Scene for Future History." Eventually ecological memory begins to reclaim every deserted farmstead through biological processes such as colonization by pioneering plants followed by subsequent succession of other native species: "there will be / fields, a hayrake, a storm, and no one / at work. . . . It will be wilderness again."[10] Ironically, he points out, "Lost pioneers are the ones / who found the best valleys"; that is, those settlers who succeeded in claiming the most bountiful land rapidly vanished from the saga of westward expansion simply by settling down to stay.[11]

In terms of cultural history, Stafford was particularly attentive to the ongoing presence—and absence—of Native Americans. Like his attunement to the natural history of the grasslands, a fascination with Indians had been forged as he grew up on the Great Plains and is reflected in poems evoking cultural memory as well as the environmental history at sites of traditional trade routes, historical migrations, and sacred landmarks. "Midwest," for instance, invokes the older order of indigenous agriculture, cultivating ancient crops in accordance with the perennial seasonal round. While subsistence cultures are subject to periodic deprivations, this poem might also be read as celebrating viability and survival, and implicitly introducing a dialectic between dispossession and the prospect of renewal:

> Corn that the starving Indians held
> all through moons of cold for seed
> and then they lost in stony ground
> the gods told them to plant it in—
> west of your city that corn still lies.[12]

In fact, Stafford's poems often demonstrate an abiding interest in the legacy of indigenous peoples, particularly their intimate acquaintance with the plants and animals that sustained them. Their symbolic significance in the popular imagination, he held, reflected the longstanding trope of associating Native Americans with natural abundance and the "pristine" wilderness they were imagined to have once inhabited, and sometimes romanticizing them as stewards of the wild.[13] Scholars such as Shepard Krech, William Cronon, and David Lentz have helped establish the environmental impact of their various subsistence activities, from setting fire to woodlands and prairies to clearing successive plots for crops. Recognizing this legacy of pre-contact agriculture, as Stafford does, tacitly acknowledges that Native peoples had long been agents of environmental change in the Americas.[14]

Cultural geographer John Hudson argues that as white settlers progressed westward—along the Ohio River and northward on the Wabash, for example—they encountered patches of land that had already effectively been altered by centuries of cultivation. These clearings and abandoned fields were suitable for European forms of farming as well as preexisting agricultural practices that were adopted by pioneers on the frontier. Since some cropland in Virginia east of the Appalachian range was already becoming exhausted by the late 1700s, settlers coveted the fertile bottomlands to the west that had been improved before their arrival:

> Aboriginal North Americans, like people everywhere, had deliberately disturbed their surroundings in order to produce an increase in nature. Burning to increase the grasses and animals who fed on them was the most important modification in terms of the area affected.... Planting was to the riverine habitats what massive burning was to uplands. It was in the bottoms that the Indians raised their crops, and there, too, would the white settlers who replaced them.[15]

Existing fields that had been abandoned could be most readily appropriated by those white settlers who arrived soon after Native relocation, while untended riparian tracts would rapidly revert to woodland. Similar themes of tribal dislocation from traditional homelands—as well as the spiritual dimensions of agricultural customs specific to ecological context—arise forcefully in Diane Glancy's *Pushing the Bear*, a pair of novels reimagining the Cherokee Trail of Tears. Stafford was an uneasy heir to the imperial project of westward expansion, settlement, and development, with its corresponding destruction of the environment and displacement of Native American nations.

Age-old agricultural practices such as cultivating corn, potatoes, beans, and squash, for example—as well as gathering nuts, roots, and berries—have long been overshadowed in the popular imagination by images of nomadic Plains tribes pursuing buffalo on horseback. Equally suspect, however, are facile stereotypes of indigenous environmentalists. As Lee Schweninger contends in *Listening to the Land*, a study of relationships with nature portrayed in Native American literature, "the concept of indigenous environmentalism [is] a romantic ideal, casting mostly pre-twentieth-century American Indians as models of ethical land stewards."[16]

Caricatures long perpetuated by Hollywood westerns to the point of cliché are prone to eclipse the prevalence and extent of agriculture in North America prior to contact with Europeans. Hunting game and planting crops were certainly not mutually exclusive activities. In fact, the two often proved to be perfectly compatible, as Barry Lopez explains: "The Pawnee were in some ways the most complex of plains tribes because they were both an agricultural and a hunting people. The renewal of their corn crops and their annual buffalo hunts were two driving forces in their ceremonials and in all of them the life-death cycle—which functioned through the agreement with the Animal Master and the renewal cycle of the crops—was central."[17] The various types of subsistence—hunting and fishing, planting and gathering—were each accompanied by complementary customs, beliefs, and rituals.

Stafford maintained that in mainstream American culture, Indians have long been perceived partially in metaphorical terms: "They stand for when the continent was unspoiled.... The mystique of it is on a shaky foundation I think, but the metaphorical significance of it is very important."[18] Native Americans situated in seemingly pristine wilderness settings is an image perpetuated for centuries by the nation's premier landscape painters. The lone figure or a band's encampment is indeed a familiar device in the works of nineteenth-century artists such as those of the Hudson River School, including Thomas Cole, Asher Durand, Thomas Moran, perhaps most famously in Albert Bierstadt's monumental canvases of the western frontier. While the term "wild Indians" has since been galvanized in the popular imagination by westerns in print and film as marauding bands perpetrating frightful raids on frontier settlements, it had once carried a markedly different connotation: where "wild" was applied to tribes not yet compromised by contact with Europeans and therefore deemed a more "pure" example of cultural traditions. While clouded by the complexities of cultural contact and hybridity, the ideal of authenticity has been widely romanticized. This was the very figure Henry David Thoreau pursued in his "Indian Notebooks"; that George Catlin portrayed on canvas as robust, ennobled, and proud; that Henry Wadsworth Longfellow enshrined in *Hiawatha*; and that was perhaps most widely promulgated since the early twentieth century by the sentimentalized photographs of Edward Curtis. Consequently, the figure of Indians in unsettled settings has long served as a conventionalized image, a socially constructed symbol suggesting an often idealized degree of pre-contact natural abundance.

Stafford's overall attunement to wildlife, assumed by some readers to suggest a Native American sensibility, is often shown in his attention to species carrying symbolic cachet. His poems sometimes connect environmental history and cultural memory explicitly by associating indigenous traditions with images of iconic animals. In "Some Shadows," from *The Rescued Year*, for example, he recounts moments from his youth on the prairies of Kansas, suggesting how those early glimpses of natural and cultural history had left an indelible impression: "When Mother was a girl Indians / shadowed that country, the barren lands / . . . Hawks cling to the barrens wherever I live."[19] Other poems such as "Lake Wendoka" portray his childhood memories of a prairie subjected to repeated landscape change in the wake of tribal relocation:

> Under the sidewalk lay an Indian village—
> . . . .
> where a hillside opened and they built a dam,
> long blue water flowing prairie lake,
> slow waves lapping, rock to bank to sand:
> *Arapaho, Kansa, Cheyenne, Cheyenne, Cheyenne.*[20]

Naming the various tribes who had made this region their homeland over millennia reads like an incantation, as if spoken by the earth itself. Yet the place being depicted had been fundamentally altered since the arrival of whites: the ruins of ancient villages paved over, watersheds erased by the construction of dams. This layering of time is characteristic of his meditations on landscapes of the past.

Stafford frequently evoked the continent's indigenous peoples in relation to specific landscapes, particularly places that historically provided habitat for species of animals with special symbolic or spiritual significance. In the Pacific Northwest, for instance, the salmon was especially revered, much like the bison of the Plains; indeed, such relationships were regarded as sacred. The sustainability of the resource—and ultimately the culture itself—entailed abiding reverently by the customs believed to perpetuate that relationship. Given the profound importance of the species to traditional lifeways, art, and religion, these anadromous fish spawning in the river of their birth provide a compelling metaphor for both ecological and cultural continuance. Stafford celebrated the marvel of their migration and spawning in *Passwords* with haiku-like concision: "Salmon return out of a wide ocean / and find their home river all the way back / through the bitter current."[21] Disruption of such ancient migrations can threaten both ecological resources and corresponding cultural memory. Besides serving as primary food sources for a variety of tribes, after all, such species carry symbolic and often totemic significance. Consequently, return of the salmon could renew a cultural as well as a subsistence resource. "Northwest Coast Indians in general recognize a critical, reciprocal relationship between the fish and the fisher specifically, and generally between the human and nonhuman community," historian Lee Schweninger explains, "moral principles that promote such a respectful and spiritual relationship."[22]

Connections between such iconic wildlife and indigenous language can be seen in how the word *chinook*—referring to a particular species of fish, the prized king salmon—also carries a complex combination of cultural connotations. *Chinook* is the name of a historic Columbia River tribe as well as intertribal trade jargon that once served as a regional *lingua franca* for traders in the Pacific Northwest: a creole of Chinook dialect mixed with Nootka, Chehalis, English, and French, as well as borrowings from several other languages. The once robust populations of salmon spawning in major western rivers have been threatened where dams have greatly truncated their previous range, with hatcheries sometimes offering the last remaining safeguard against catastrophic decline. Anadromous species such as salmon that spawn in freshwater had historically supported runs reaching deep inland on watersheds such as the Snake, Columbia, and Yukon, as well as a myriad of smaller streams. Stafford laments dams obstructing such ancient migrations in "The Fish Counter at Bonneville," from his first collection, *West of Your City*, published in 1960: "Downstream they have killed the river and built a dam . . . / the ghosts of game . . . / So many Chinook souls, so many Silverside."[23] Salmon were once the very lifeblood of many Native nations along these rivers and their tributaries. Consequently, the obstruction by massive dams of major rivers connected to vast networks of salmon streams has threatened to undermine age-old subsistence relationships throughout the twentieth century. Stafford's poem "Fish Counter" implicitly conveys "the Native American point of view, since the Columbia River tribes lost their ancient village sites and fishing rocks had to be relocated due to the dam's construction."[24] Damming such waterways for power stations has inevitably risked disrupting migrations extending deep into geological time, thwarting spawning instincts that have sustained these species since at least the last Ice Age.

Stafford clearly recognized how major predators such as wolf, bear, cougar, and eagle often held elevated cultural as well as environmental significance. Traditionally these creatures were often seen as embodying virtues human hunters aspired to, such as stealth, endurance, strength, and vision. As Lopez notes, many Native American peoples associated the cardinal directions with totemic animals such as bear, wolf, wildcat, and mountain lion (West, East, South, and North, respectively) because these were "creatures with the greatest power and influence in the world."[25] Those four animals, along with the eagle, were viewed not only as great hunters but also assigned the highest cosmological significance. The same animals held in totemic regard by many indigenous peoples—including those that have been subject for centuries to extirpation and systematic eradication throughout North America—often correspond to those that contemporary ecologists consider to be indicator or even keystone species. Such apex predators maintain healthy populations of prey species and thereby preserve the integrity and vitality of entire ecosystems, relationships initially posited by pioneering ecologists and wildlife biologists such as Aldo Leopold and Paul Errington.

Yet for Stafford, the buffalo was a quintessential emblem of environmental history on the continent. His poems repeatedly invoke the legendary herds and their once seemingly unimaginable abundance, hunted to the very brink of extinction, only to be revived as livestock in our time. He realized that portions of their far-flung historical range are sometimes still traced by today's highways, which were preceded by rutted wagon train roads, which in turn had followed Indian trails, and which originally had been buffalo traces. As Hudson explains, "The bison themselves were agents of change... [that] left their mark on the land... an unmistakable network of good trails (traces) that first the Indians and then Europeans traveled.... Wherever the bison were, there were found modifications"—to the environment, disseminating seeds, tearing up turf where they once roamed, traces that remarkably still exist in many places today, more than a century after the decimation of the herds.[26]

Age-old buffalo traces were also directly linked to westward migration, being followed by pioneering homesteaders since they invariably offered passable routes through the most challenging terrain. Until the late eighteenth century, in fact, thoroughfares for wagons in eastern states such as Kentucky still retraced trails bison had blazed: "In addition to acting as compass lines for colonial migration," Tom McHugh observes in *The Time of the Buffalo*, "the trails influenced the patterns of settlement in the heart of the Ohio River valley," accounting for the location of today's population centers such as Cincinnati and Louisville. "Although the buffalo disappeared soon afterward from most of this area east of the Mississippi,"

McHugh concludes, "the herds had already stamped the country with some distinctive signs of occupation that would persist for centuries."[27] As cultural geographer Richard Manning observes in *Grassland*, his seminal account of America's prairies, "A single bison does not stand alone, is not an individual. It is, rather, a manifestation of a place, the net result, the capstone of fire, wind, and grass—grass to the horizon—and of the hundreds of plants that live on it.... And can only be understood as such."[28] Stafford's poems evoke just such ecological, historical, and cultural legacies:

> Comanches tell how the buffalo
> wore down their own pass through these hills,
> herds pouring over for years, not finding
> a way but making it by going there...
> ...I bow my head
> in the graveyard at Buffalo Gap.[29]

The poet pays homage, remembering the abundance, the loss, and above all the stories.

Similarly, he characterizes a frontier cemetery in South Dakota in terms of environmental history, acknowledging the role of buffalo in shaping the prairie biome: "to know the world as a land invented / by breath / its hills and plains guided / and anchored in place by thought, by feet."[30] Yet he also contemplates the brink of their extinction in the nineteenth century in "By a River in the Osage Country," a poem again establishing place-based connections between environmental history and cultural memory. The river itself is named "Neosho," he tells us, meaning "a river made muddy by buffalo."[31] The poem offers a lament yet also seems to hold out hope for their return and restoration: a living legacy, a symbol of both vulnerability and resilience, an emblem of the continent's once seemingly inexhaustible abundance. "By a River" relates the poet's own experience visiting the place:

> ...I have waded that river
> and looked into the eyes of buffalo
> that were standing and gazing far:
> no soul I have met knew the source
> that well, or where the Neosho
> went when it was clear.[32]

Water once muddied by hooves now ran clear, reflecting the precipitous decline of the herds that had once crossed there. Indian nations that had long inhabited or hunted in the region—the very peoples who named the river "Neosho"—suffered a similar fate of dispersal and dislocation at the onset of the reservation era. The river is significant in the saga of westward settlement as well. Council Grove on the banks of the Neosho was a major staging area for caravans of prairie schooners before they headed out for western territories. These were the storied wagon trains of yore that crossed the High Plains along the Arkansas River before veering south to Santa Fe.[33] "By a River" provides a striking reminder of how at that time declining wildlife coincided with Indian removal in the name of Manifest Destiny and ultimately white settlement. Here, reading environmental history entails accounting for cultural memory as well.

For Stafford, the decline of the buffalo was intertwined with the travesty of Indian removal. In "A Sound from the Earth," originally published in the collection *Allegiances*, he recounts the unearthing of buffalo bones long after the decimation of the herds:

> Somewhere, I think in Dakota,
> they found the leg bones—just the
> big leg bones—of several hundred
> buffalo in a gravel pit.
> ....
> The grandfather of Crazy Horse
> lived there...[34]

Referring to Crazy Horse in this context ties the near extinction of the buffalo to the final years of Plains Indian resistance. It also evokes the totemic power—or medicine—ascribed to the buffalo by invoking the legendary leader whose own grandfather chides:

> where was your buffalo medicine?
> "I say you were not brave enough, Boy.
> I say Crazy Horse was too cautious."
>
> Then the sound he cried out for his grandson
> made that thin Agency soup that they
> put before him tremble.[35]

"Report to Crazy Horse" offers a particularly poignant glimpse of that Lakota history. By addressing a major figure from the Indian Wars of the nineteenth century, this poem suggests an enduring spirit of resistance and resilience. In the voice of a survivor faced with assimilation, Stafford effectively juxtaposes the "vision" of tribal leaders with the ascendant Anglo order. It portrays ambiguity and uncertainty on the cusp of two cultures:

> Crazy Horse, it is not fair
> to hide a new vision from you.
> In our school we are learning
> to take aim when we talk, and we have
> found our enemies....
> ....
> ...A teacher here says
> hurt and scorned people are places
> where real enemies hide....
> ....
> [I]n a brave way, the way Crazy Horse
> talked: that teacher is right.[36]

Originally published in *Someday, Maybe*, this poem undoubtedly contributed to the positive reception Stafford met when reading his work in Native American communities. "When he read his poems in North Dakota on one occasion, many Sioux people came to listen to what he had to say about Crazy Horse," as his son Kim recounts.[37] In *Early Morning*, he offers this account of the event: "Once after returning from a series of poetry readings in the Dakotas, my father said many in the audience had been Lakota people. They had heard he would be reading poems about Crazy Horse. There was a feeling on the Pine Ridge reservation then that Crazy Horse might return."[38] Being welcomed by this predominantly Lakota audience at Pine Ridge signaled shared hope for cultural revival. Indeed, "Report" closes with a heartening appraisal of resources the land still affords traditional subsistence peoples:

> The chokecherries along our valley
> still bear a bright fruit. There is good
> pottery clay north of here. I remember
> our old places. When I pass the Musselshell
> I run my hand along those old grooves in the rock.[39]

Above all, Stafford recognized that the near extinction of the buffalo had become tragically intertwined with the fate of the Native Americans during the nineteenth century. Much like the wholesale destruction of habitat that has all too often destabilized subsistence cultures, the decimation of the buffalo herds by the 1880s caused starvation among those peoples dependent on them. Plains Indian cultures in particular were completely undermined by extermination of the bison. Consider the case of the Osage. *Talking to the Moon*, the autobiography of John Joseph Mathews, chronicles the unexpectedly sudden reduction of the buffalo's historic range, including tribal hunting grounds along the Cimarron in Oklahoma, by the onslaught of market hunters and then cattle ranchers. The fate of the Osage Nation seems emblematic of the way the traditional subsistence practices of Native peoples have been jeopardized in region after region by successive waves of development and destructive landscape change.

The near extinction and subsequent revival of the buffalo has also become a founding narrative of the country's early conservation movement, associated with the establishment of the first national parks and precursor to a host of preservation and restoration initiatives to the present day. Ioway historian Lance Foster plainly equates the two: "The buffalo and the prairie-plains were synonymous. One cannot truly and deeply exist without the other. While the prairies sustained the buffalo, the buffalo maintained the prairie.... The buffalo shaped the landscape of the prairie in many ways."[40] Stafford himself recognized the pivotal ecological role the herds once played on the Great Plains. His poem "Before It Burned Over: A Sioux Grass Chant" opens by invoking the profound ecological importance historically in creating and maintaining the prairie biome: "World carpet, robe, every leaf / to look at, brought forth by / buffalo, green places maker."[41]

For this reason, contemporary wildlife ecologists such as Jonathan Proctor lament the near absence of free-roaming herds on the Great Plains today: "Bison are keystone species. This means their presence—or absence—alters the ecosystem. Migrating, grazing bison once created habitat for prairie dogs, grassland birds, and even many plant species."[42] He contends that despite their resurgence as commercial livestock, they are extinct in an ecological sense. Accordingly, attempts to restore portions of the prairie biome are in part contingent on successful reintroduction of this keystone species that historically helped shape that landscape. The Nature Conservancy's thirty-six-thousand-acre Tallgrass Prairie Reserve, for instance, is predicated on reintroducing buffalo, as Richard Manning writes in *Grassland*: "The conservancy is using the majesty of bison to fire some national imagination about the majesty of grass.... Saying the bison are back is only another way of saying the prairie is back.... All of the grassland West is a candidate for restoration ecology, a candidate for resurrection, and that resurrection must include bison."[43]

To place such restoration initiatives in perspective, consider a fundamental shift in the thinking of conservationist biologists over the last half-century. Progressive wildlife managers speak now in terms of eco-regional plans to protect "key tracts of nature collectively representing the whole, whose protection would conserve the region's native species...portfolios of preserves, those networks of refuges that would conserve the greatest breadth of biodiversity for the long term."[44] For instance, the protection of an unprecedented 67 million acres by the Forest Service and Bureau of Land Management in 2015, to ensure the survival of the greater sage-grouse—and with it the habitat of 350 additional species of wildlife native to the sagebrush ecosystem—is completely in keeping with this emerging bioregional paradigm. As renowned wildlife biologist George Schaller explains: "Instead of focusing just on discrete, isolated protected areas, conservation has enlarged its vision to manage whole landscapes. The goal is to create a mosaic of core areas without people or development...connected by corridors of viable habitat."[45]

Other contemporary wildlife conservation initiatives such as the American Prairie Reserve seek to preserve critical tracts of habitat corresponding to historical range as well as contiguous corridors for unimpeded migration. As Manning points out, "the grass can live again. Unlike forest, it recovers rapidly, some of it in a matter of five to ten years. There is a possibility of resurrection in a real and physical sense. There is the possibility of bison and wolves."[46] In short, to restore the ecological integrity of grassland ultimately entails reintroducing the species that historically helped shape the prairie biome in its entirety. While Stafford's writing predates such developments in conservation biology, he intuitively grasped the relationship between the integrity of ecosystems that support such keystone species and the cultural legacy of peoples who had so long relied on them for material sustenance as well as indigenous identity. Revival of the bison in our time implicitly signals the prospect of cultural renewal.

With his frequent allusions to Indian culture and history, it is little wonder that Stafford was occasionally mistaken for being of Native American descent himself. Dozens of poems in *The Way It Is*, a volume of new and selected work published in 1998, refer to a variety of Indian peoples, including the Apache, Blackfoot, Comanche, Cree, Klickitat, Modoc, Navajo, Nez Perce, Osage, Sioux, Snake, Umpqua, and Yakima, as well as significant

Indian figures such as Chief Joseph, Crazy Horse, Ishi, and Wovoka. Those who misconstrue Stafford's work as "Indian" writing perpetuate the assumption that sensibilities stereotypically associated with cultural identity can be attributed automatically to an author's personal ancestry. In truth, it is the poems, rather than the poet himself, that reflect indigenous values. As Ojibwe author, linguist, and critic David Treuer has observed, "Who is and who isn't an Indian is a complicated question to answer, but there are many ways to answer it beyond genetics alone.... Bending to a common purpose is more important than arising from a common place."[47] In Stafford's case, common purpose would include values such as reverence for the natural world and adherence to ethical relationships with every living thing.

The presence of Indian material in his poetry, according to pioneering ecocritic Lars Nordstrom, ultimately serves as an "emblem of an attitude toward wilderness." In this formulation, "Stafford's poems attempt to articulate how we are to live correctly and responsibly within the natural order."[48] Perhaps what matters most is cultural *perspective*. Those critics who have focused narrowly on Stafford's ancestry appear to have missed a deeper truth: the poet's enduring interest in indigenous perspectives ultimately led him to view landscape through the lens of cultural memory and environmental history. The diversity of Indian cultures, he realized, derives in part from the continent's great range of ecosystems, an assessment reminiscent of Aldo Leopold's claim that each culture is ultimately shaped by its perceptions of nature, and the premise of Louise Erdrich's *Books and Islands* that every indigenous language is connected inextricably to the place it evolved. As Donelle Dreese concludes, "the diversity of the landscape of what we now call North America has contributed greatly in creating a diversity of American Indian cultures."[49]

Time afield observing wildlife across the continent provided the occasion for Stafford to meditate on ancient and enduring aspects of that relationship. Witnessing the annual arrival of sandhill cranes, he contemplated why such great migrations stir the human imagination, as pioneering ecologists and literary naturalists such as Aldo Leopold, Paul Errington, and Paul Gruchow have done. Stafford characterized the birds as "travelers" and "far wanderers," emissaries from deep time. "They extend our life," he maintained, "piercing through space to reappear / quietly, undeniably, where we are," riveting us in a present stitched to the distant past.[50] "Birds are a hope," he proclaimed in *Sound of the Ax*, a posthumous volume of epigrammatic adages and aphoristic verse, "they can find the islands left."[51]

Accordingly, international efforts to protect songbirds whose migrations span several continents must consider the integrity of both breeding habitat in North America and winter range in tropical forests, as well as areas for feeding on the epic flight between the two: stopover spots for flocks to forage en route. For Stafford, the order of magnitude involved—the abundance of birds flocking in almost unimaginable numbers, as well as the vast distances being crossed—suggested to him a magnificent primal mystery:

> ... farther than winter,
> over the town a coming of birds.
> What great wild hands will reach for them?
> —*and for all who are here when those wanderers come?*[52]

Similarly, in "Spirit of Place: Great Blue Heron," he takes particular solace in encountering such an ancient species "at the edge" of civilization: "It is / the wilderness come back again," he proclaims, to "a lagoon / with our city reflected in its eye. / We live by faith in such presences."[53] The annual recurrence of their migratory flights seems to offer a sort of reassurance as it reasserts an ancient continuity, providing a poignant reminder that even given the immensity of evolutionary time, such migrations endure, and with them the perpetuation of fundamental ecological processes linked to the cycle of seasons.

Stafford's poems depicting myriad landscapes across the continent reflect the geographic trajectory of his life, spanning an upbringing on the grasslands of Kansas, subsequent travels throughout North America, and ultimately residence in the Pacific Northwest. Whether evoked by the buffalo fording the Neosho or salmon returning to spawn on the majestic Columbia River, his reading of each landscape is grounded in a personal reverence as well as an appreciation for cultural history. In the end, his deep-seated affinity for indigenous peoples was predicated on shared concerns for ethical relations with the natural world, and his conception of place was situated at the cusp of environmental history and cultural memory. Yet he also recognized that our perceptions of the living earth around us are continuously reshaped over time, even as landscapes themselves have changed. "The stream itself," he reminds us, "is always revising."[54]

NOTES

1. Stafford, *The Way It Is*, 60–61.
2. Stafford, *Stories That Could Be True*, 29.

3. Stafford, *Sound of the Ax*, 31.

4. Stafford, "The Peters Family," *Stories That Could Be True*, 97.

5. According to editor Fred Marchant, "White Pigeons" was "his first poem, as indicated by Stafford's own notation on the typescript," which reads, in part, "the last semester [as an undergraduate at Kansas University]. The first time I really tried to express poetry," as cited in Stafford, *Another World Instead*, xv, 137.

6. Stafford, *Another World Instead*, 7.

7. Nye, introduction to *The Way It Is*, xx.

8. Stafford, "By the Snake River," *Stories That Could Be True*, 43.

9. See Jonathan Raban's *Bad Lands* as well as Ian Frazier's *Great Plains* for an account of initial settlement of the region and subsequent departure of pioneers in the wake of generations of failed attempts at farming.

10. Stafford, "Roadside Markers for West of Dodge," *Kansas Poems*, 9.

11. Stafford, *Sound of the Axe*, 1.

12. Stafford, *Stories That Could Be True*, 29.

13. See *The Ecological Indian* (2000) by Shepard Krech and *Listening to the Land* (2008) by Lee Schweninger for a thorough treatment and critique of such stereotypes.

14. The impact of such indigenous agricultural practices on the landscape is explored by John Hudson in *Making the Corn Belt: A Geographical History of Middle-Western Agriculture*.

15. Ibid., 43–44.

16. Schweninger, *Listening to the Land*, 26.

17. Lopez, *Of Wolves and Men*, 132.

18. Stafford, "Willingly Local."

19. Stafford, "Some Shadows," *The Way It Is*, 103.

20. Stafford, *Kansas Poems*, 25.

21. Stafford, "Different Things," *The Way It Is*, 241.

22. Schweninger, *Listening to the Land*, 117.

23. Stafford, *Stories That Could Be True*, 43.

24. Nordstrom, *Theodore Roethke*, 61.

25. Lopez, *Of Wolves and Men*, 102.

26. Hudson, *Making the Corn Belt*, 26.

27. McHugh, *Time of the Buffalo*, 49.

28. Manning, *Grassland*, 245–46.

29. Stafford, "*Paso por Aquí*," *The Way It Is*, 235.

30. Ibid.

31. Ibid.

32. Ibid., 229.

33. Hudson, *Across This Land*, 258.

34. Stafford, *The Way It Is*, 124.

35. Ibid.

36. Stafford, *The Way It Is*, 139.

37. Kim Stafford, e-mail message to author, August 15, 2009.

38. Kim Stafford, *Early Morning*, 80.

39. Stafford, *The Way It Is*, 140.

40. Foster, "Tanji na Che."

41. Stafford, *The Way It Is*, 181.
42. Proctor, "Defender to the Core," *Defenders* 86, no. 4 (Washington, DC: Defenders of Wildlife, 2011), 13.
43. Manning, *Grassland*, 217.
44. Stolzenburg, "Long Rangers," 41.
45. Schaller, "Politics Is Killing the Big Cats," 90.
46. Manning, *Grassland*, 9.
47. Treuer, "How Do You Prove."
48. Nordstrom, *Theodore Roethke*, 62.
49. Dreese, *Ecocriticism*, 8.
50. Stafford, *The Way It Is*, 173.
51. Stafford, *Sound of the Axe*, 4.
52. Stafford, "Summer Will Rise," *Stories That Could Be True*, 85.
53. Stafford, *The Way It Is*, 167.
54. Stafford, *Sound of the Axe*, 67.

7

Landscape and Language

Louise Erdrich (b. 1954)

We have been shaped not only by our human ancestors but also by the environments in which they lived.

—Joan Halifax, *The Fruitful Darkness*

We are words on a journey
not the inscriptions of settled people.

—W. S. Merwin, "An Encampment at Morning"

Part of the trip is always the return.

—Louise Erdrich, *Books and Islands in Ojibwe Country*

Louise Erdrich is one of the most accomplished and highly esteemed of contemporary Native American writers. The National Book Award she received in 2012 for *The Roundhouse* confirmed her place among the nation's major novelists. This chapter addresses a work especially important to her ongoing exploration of identity in relation to her Ojibwe heritage, *Books and Islands in Ojibwe Country*, her second memoir, published in 2003. Her account of a canoe trip on the Boundary Waters of southern Ontario, Canada, known as Lake of the Woods, is essentially autobiographical, although it might equally be termed *ecocultural*.[1] Erdrich contemplates the relationship between natural and *cultural* resources, pondering how Ojibwe language and tradition are linked to this particular landscape; by acquiring a language native to the region she traverses, she reflects on how culture intersects directly with the natural world. Composing this work, much like making the journey itself, constitutes not only an exploration but a *restoration* of an important portion of her cultural heritage, a localized mode of knowing, involving an ancient and enduring attunement to place. In this literary travelogue she bears witness to indigenous history—particularly the ancient rock paintings left by ancestral Anishinaabe.

The trip Erdrich recounts is expressly predicated on language revitalization, paddling through the Boundary Waters with a group dedicated to studying Ojibwemowin. She had initially been tantalized by listening to fluent elders on Manitoulin Island, Ontario, and then resolved to learn to speak it herself. On this journey to Lake of the Woods, she seeks to further her study in the place the language evolved, visiting the region with members of the Ojibwe Language Society, a group that meets weekly at Lac Court Oreilles. Traveling with her infant daughter and the girl's father, Tobasonakwut, who also serves as her mentor and guide, she describes the experience in terms of cultural, linguistic, and, importantly, *natural* history, learning words in Ojibwemowin that correspond to the landscape of these ancestral homelands. They set out for an isolated library of Native American historical and cultural documents improbably housed in a remote wilderness setting, where this group of teachers and serious students of Ojibwemowin will gather to practice. Erdrich hopes to engage in linguistic reclamation herself by learning to speak the language as a source of cultural and ultimately ecological memory. In the process, she discovers how indigenous languages are inextricably linked to particular landscapes and ecosystems.

Books and Islands depicts the trip as a kind of pilgrimage, expressing veneration for the natural world in conjunction with indigenous language and culture. Places held sacred by the Ojibwe, including the locations of ancient rock paintings dating back hundreds—even *thousands*—of years, remain in living memory to this day. While clearings for traditional villages and Grand Medicine society, or Midewiwin, lodges, for instance, have been swiftly reclaimed by a succession of northern woodland trees, within a single generation of being abandoned shortly after World War II, those locations are still known to the Ojibwe tradition bearers, such as Tobasonakwut, among the last to have been born and raised at Lake of the Woods. Above all, Erdrich senses an enduring connection between this Native American language and the natural world in the region where it originated.

In addition to contemplating the origin of native languages in relation to place, Erdrich describes being utterly enthralled by the ancient rock paintings, known as *mazinapikiniganan* to the Ojibwe, found throughout much of the Boundary Waters region. The pictomythic glyphs in the rock paintings she encounters—each a repository of cultural memory—correspond to traditional songs that often incorporate images derived from visions and dreams, and so represent a microcosm of indigenous cosmology. She recognizes that these images painted on the islands in Ojibwe country trace a sacred relationship to nature that reflects an ethic that sustained a people. This history deepens her desire to discern the remaining traces of an ancient legacy still embodied by Ojibwemowin and still intelligible today.

Erdrich and her party traverse part of a park that straddles the border between the United States and Canada. The location underscores the history of colonization whereby the Ojibwe—like most other Indian nations—were dispossessed of their ancestral territories. Erdrich's personal quest to recover traditional culture and language might be likened to a kind of border crossing, with vast temporal as well as topographic dimensions. A map of "Ojibwe Country" appears at the beginning of her account, harkening back to narratives from the age of exploration, and reminding us once more of its colonial consequences. By the seventeenth century French voyageurs had already plied major rivers in the region, accompanied by a lethal amalgam of epidemics, alcohol, and firearms. The Ojibwe participated in the fur trade from its inception, and this, along with successive waves of Christian missionaries, undermined traditional beliefs predicated on a sacred relationship to nature and reverence for game.

The trip retraces a path taken by explorers and traders, trappers and priests—as well as soldiers. In fact, Erdrich notes, the road taken to reach Lake of the Woods has an even deeper history, as a traditional Ojibwe trade route: "Songs traveled this route, and ceremonies, as well as pelts and guns. Medicines, knowledge, sacred shells, and secular ideas traveled this road."[2] Such reflections invoke cartographic histories, including maps produced by early explorers that rendered the verdant New World in two dimensions, conducive to articulating territorial claims by European—and ultimately—American nation-states. In a sense, this journey she has taken back through time and space, as well as her narrative of that trip, reclaims the traditional territory of her Anishinaabe ancestors, and with it the Ojibwe cultural heritage that had come into being in relation to this landscape. The archaeological record suggests that this area was their ancestral homeland for at least three millennia, although Ojibwe tradition holds that their people have in effect *always* lived here. Such a quest necessarily entails a journey of *return*.

Erdrich clearly cherishes her connection to Ojibwe history, especially indigenous language and its relationship to the environment in the region where it evolved. Writing this book in conjunction with the expedition becomes the occasion for profoundly personal and philosophical reflections on cultural identity and its relation to nature. This journey in service of language learning corresponds to what David Treuer, an Ojibwe linguist, literary critic, and novelist from the Leech Lake Reservation in northern Minnesota, terms "an idea of culture that functions as memory."[3] Speaking Ojibwemowin herself provides a vital link to that Native American heritage she believes is essential to understanding her own place in the world.

Indigenous language has always represented for Erdrich an especially precious vessel for preserving and perpetuating traditional knowledge and values, perhaps above all as an embodiment of an ancient cultural relationship to nature and place. Many of her novels have incorporated Ojibwemowin as a marker of tradition and identity, often suggesting a yearning for the connection that language provides to cultural heritage. "The longing for culture is linked to the project of self-recovery and self-discovery," Treuer observes. "The importance of 'the old language' is the most obvious aspect of cultural longing in Erdrich's writing."[4] He suggests that in her masterful first novel, *Love Medicine*, for example, "the use of Ojibwe words . . . highlights the longing for culture."[5] "It is a question of culture expressed as sentiment, as a wish."[6] Such a desire for "cultural reconnection," he points out, recurs in major works by several prominent Indian writers, in which indigenous languages signify traditional identity, such as Momaday's *House Made of Dawn*, Silko's *Ceremony*, and Welch's *Winter in the Blood*. Similarly, in Erdrich's *Love Medicine* characters "reflect on culture in long, uninterrupted interior monologues, and such reflections are typically cast as reflection on Ojibwe language."[7] In *Books and Islands* the author *herself*, rather than a character, is moved to acquire Ojibwemowin. "There is an urgency," she discovers, "about attempting to speak the language."[8] Indeed, in an era when many indigenous languages are threatened—or already lost—this one is still being spoken.

Erdrich recounts how the Anishinaabe language has long been associated geographically with traditional Ojibwe country. "Ojibwemowin is the Algonquin language originally spoken by the Ojibwe people living throughout Michigan, Minnesota, Ontario, Manitoba, and on into North Dakota."[9] Historian Ute Lischke plots a sweeping geographical map for the region's Indian peoples: "The Chippewa nation has resided since time immemorial in birchbark country, the country along the Great Lakes and connecting waterways of northeastern North America, as far west as present-day North Dakota and Manitoba. The Cree and Metis nations continue to reside over the Prairie provinces and northern Ontario and Quebec in Canada, as well as in the northwestern United States."[10] In fact, the Algonquin language family as a whole once encompassed much of eastern North America, as linguist John Nichols describes: "At the time of the European invasion of North America, the languages of the Algonquian language family were spoken by Indians along the Atlantic coast from what is now North Carolina to Newfoundland, inland across Canada to the Great Plains, and in the region of the Great Lakes, perhaps ranging as far south as Alabama and Georgia."[11]

Even after withstanding centuries of linguistic assimilation, Ojibwemowin thrives in communities in regions such as northern Manitoba and Ontario to the present day. In her own lifetime, Erdrich explains, "most Ojibwe people on reserves in Canada, and many in Minnesota and Wisconsin, still spoke English as a second language, Ojibwemowin as their first."[12] Lischke characterizes her complex ethnic inheritance as "Metis/Cree/Chippewa (also known as Anishanaabe of Ojibwa) heritage on her mother's side and German/Jewish/Catholic on her father's side."[13] Erdrich describes her own maternal grandmother's generation as possessing a similarly multifaceted ethnic ancestry: "Ojibwa, and Scots [and certainly Metis/Cree]."[14] In her family, Ojibwemowin remained the first and primary language up until two or three generations ago. As the daughter of an Ojibwe-French mother, Erdrich was raised in North Dakota, where she maintains ties to the Turtle Mountain band to this day. Her grandfather, she explains, "was the last person in [her] family who spoke his native language, Ojibwemowin, with any fluency."[15]

Erdrich contrasts her yearning to learn the language with the linguistic assimilation faced by native speakers of previous generations. "My experience with the language is of course very different. Instead of the language being beaten out of me, I've tried for years to acquire it. But how do I go back to a language I never had."[16] We repeatedly sense the author's desire to reembrace this culture; the impulse that has led her here reflects an underlying hunger to engage the language. She recalls that "growing up in an ordinary small North Dakota town, I thought Ojibwemowin was a language of prayers."[17] Now mindful of how the traditions of Native Americans have so often been romanticized, she concedes that while it "is a language that most directly reflects a human involvement with the spirit of the land itself...that is not to say Ojibwemowin is an elevated language of *vanished* spirituality."[18] Admittedly, her aspiration to gain greater fluency in this *living* Indian language differs fundamentally from attempts to restore those Native tongues no longer spoken today, as well as those predicted to meet extinction within a single generation.

Still, Ojibwemowin is a language of memory, Erdrich reminds us, likening stories from oral tradition to "books in the people."[19] Ojibwemowin, she observes, "is one the few surviving languages that evolved to the present here in North America...a language that most directly reflects a human involvement with the spirit of the land itself. It is the language," she concludes, "of the paintings."[20] *Books and Islands* articulates an even more daring claim: Ojibwemowin—and, by extension, Native American languages generally—are the linguistic repository of ecological memory, reflecting the particular places and landscapes where they evolved, as well as exemplifying a culturally mediated relationship with nature. What might indigenous languages, particularly with their place-based lexicons, reveal about the relationship between cultural and environmental history?

Consider how species biodiversity and the survival of indigenous languages sometimes seem to coincide geographically, calling to mind E. O. Wilson's theory of island biogeography, that groundbreaking and influential hypothesis positing that species biodiversity can be predicted mathematically based on two factors: the size of an island and its distance from the nearest major landmass, such as the mainland. While initially applied to maritime environments, the theory has also proven to have explanatory power in other geographical settings, such as landlocked "islands" of intact habitat surrounded by a veritable sea of human use, as opposed to actual ocean. Could a similar principle account for the survival of indigenous languages? Imagine if our intuitive association of cultural and environmental change—and continuity—were grounded in the fundamental fact of their coevolution.

A causal relationship might plausibly be predicated on the ecological integrity of large and isolated tracts of habitat that support natural abundance at or approaching historical levels, still sufficient to sustain age-old subsistence cultures, enabling the perpetuation of traditional languages and customs. In this light, the difficulty of preserving threatened and endangered Native languages might even be linked to the uncertain prospects for ecological restoration. As contemporary ecologists now concede, what once seemed the absolute environmental and cultural isolation of the world's most remote locations has in our time effectively ceased to exist anywhere on Earth. In fact, many environmentalists have begun to reassess the aims of landscape restoration, shifting from reconstituting historical suites of species to protecting important ecological *processes,* especially when it is no longer possible to redeem what has been lost.

Knowledge of specific habitats and ecosystems in all their particularity matters greatly to traditional subsistence cultures. As Aldo Leopold ventured in the 1949 essay "Wilderness," "the rich diversity of the world's cultures reflects a corresponding diversity in the wilds that gave them birth."[21] Accordingly, he reasons, one compelling argument for preserving wilderness beyond any biological imperative is to allow further "study of the origins of their cultural inheritance."[22] In *Books and Islands,* Erdrich in effect embarks on just such an ecocultural investigation. Her journey of linguistic and cultural restoration occurs amid myriad islands with ancient paintings on their rock faces that form in aggregate a mosaic composed of a host of ecologically significant species. Critic David McNab recounts that when Erdrich "visits Garden Island, Big Island, Painted Island, Mirage Island, Massacre Island, and Whitefish Bay," she discovers "the paintings are 'alive' and a key to the 'spiritual life' of the Ojibwe," *alive* precisely because the practice of making offerings at such sacred sites survives.[23]

In many respects, the book belongs to a tradition of American nature writing imbued with ecological awareness. "Environmental texts," as Lawrence Buell maintains, "practice restorationism by calling places into being, that is, not just by naming objects but dramatizing in the process *how they matter.*"[24] Moreover, this work explores a number of themes central to ecocriticism since its inception, such as "reciprocal relationships between humans and land" and "interconnections between nature and culture."[25] *Books and Islands* also embodies several of the premises of "environmentally oriented works" articulated by Buell in *The Environmental Imagination,* and perhaps above all a fundamental tenet: "that human history is implicated in natural history."[26]

Books and Islands also belongs to a body of writings about nature based on *journeys* that Slovic characterizes this way: "Many literary naturalists imitate the notebooks of scientific naturalists, the logbooks of explorers, or even the journals of nonscientific *travellers.*"[27] The trope of the journey frequently underlies much natural history writing, of course. Take Thoreau, for example. Critic William Rossi attributes the very "development of Thoreau's environmental imagination and natural history writing" to *travel,* specifically his journey from Walden Pond to explore the forests of Maine.[28] Rossi maintains that for Thoreau, "writing a narrative in his journal of this first wilderness" precipitated "a radical reconception of humanity's place in nature."[29] Even Thoreau's essays, he contends, can be viewed as "narrated from the perspective of the returned traveller."[30] Ultimately, Thoreau's account in *Maine Woods* helped establish the trope of the journey as a staple of American nature writing. The chronological and episodic structure of *Books and Islands* harkens back to such precursors. While contemporary American "nature writers" such as Wendell Berry trace their *return* to a place after several generations, and are largely concerned with ecological ethics—what Slovic characterizes as "processes of reclamation, rehabilitation"—the historical dimensions of Erdrich's "return" span *millennia.*[31]

Buell has argued that the trope of "(re)turn from the city to the rural place of cultural origin and spiritual centeredness" has become a conventional one in contemporary Native American writing, exemplified by a host of authors, including Momaday and Silko as well as Simon Ortiz.[32] Critics such as McNab have amplified this claim: "Returning home to Indigenous places is a *major* theme in the writings of Native American writers."[33] Yet why exactly do "indigenous places" possess such cultural relevance—and *resonance*? Her "return" to traditional Ojibwe homelands at Lake of the Woods leads Erdrich to reflect on just such questions. Admittedly, the very idea of "returning" to Indian communities from distant cities implies the loss of *ongoing* contact with ancestral landscapes. Nelson emphasizes the importance of place itself, since "in many...works of American Indian literature, both traditional and recent, identity with a physical landscape precedes cultural re-entry."[34]

Erdrich is more often portrayed and perceived in terms of her Indian cultural heritage than her relation to region or place per se, despite deep, lifelong ties to the Upper Midwest.[35] Yet as Nancy Bunge, a scholar of midwestern literature argues, appreciating Erdrich entails recognizing relation to *place* since her writing "has a strain of persistent mysticism born of a salutary unity with nature. When Erdrich's work recalls this element, it evokes not only Ojibwe roots, but Midwestern ones."[36] Indeed, as Lischke suggests, "Erdrich has often been seen as a Native American writer and as someone with local roots...'teller of the stories of her people' from North Dakota...[and] the border country between Canada and the United States."[37] Yet he also points to the geographic reach of her subject matter and readership, hailing Erdrich as "an international writer whose work encompasses and transcends North Dakota, Europe, and Turtle Island itself."[38] Still, as McNab concludes, "places and especially the Canadian connections, have been significant in Erdrich's life, her stories and writings. Nature is landscape and mindscape conjoined for Indigenous people."[39]

Connections to place figure prominently, of course, in contemporary Native American literature. Robert Nelson observes that the protagonists in major novels—such as N. Scott Momaday's *House Made of Dawn*, Leslie Marmon Silko's *Ceremony*, and James Welch's *The Death of Jim Loney*—effectively define their identity by "finding ways—sometimes 'traditional,' sometimes 'innovative,' and sometimes a creative blend of the two—to live *with* the land, holding and being held by the life that precedes and survives the life of any individual as well as culture."[40] In Erdrich's writing particularly, Lischke suggests, "the land becomes a signpost for how all the characters interact, how they are connected to each other and the land."[41] Moreover, Nelson observes, in fiction by Indian authors, places effectively function as "principal characters" in the narrative, transcending their conventional role as setting alone. He draws a fundamental distinction between the perspectives of European settlers and those of indigenous Americans whose subsistence traditions involved direct engagement with the natural landscape, in a relationship predicated on ethical principles of reciprocity. Given the profound significance of place that critics have long identified in major works of fiction by Native American authors, why not consider the ways in which *place* operates in a work of nonfiction such as *Books and Islands*?

In *Place and Vision*, Nelson posits the "function" of place in relation to Native American cultural identity and its representation in literature. He argues that for Indian individuals and communities, "a key element of 'cultural literacy'...is an intimate knowledge of a particular landscape... particular landscapes that precede and at least partially determine the cultural ones."[42] According to Lischke, "Louise Erdrich's attachment to place and land is a manifestation of her Metis roots. She has commented on place in many of her writings.... For her place...is everything."[43] Accordingly, her account of the journey to Lake of the Woods centers on how language and tradition embody a cultural legacy in relation to the ecosystems of historic Ojibwe homelands. In this formulation, place forms the bedrock of cultural—and ultimately individual—identity. For Ojibwemowin, the aspiration on which her quest squarely rests evolved in this particular environment.

The cultural legacy encoded in indigenous language arguably parallels the biological legacy defined in Leopold's seminal essay "Conservation Esthetic": namely, "the perception of the natural processes by which the land and the living things upon it have achieved their characteristic forms (evolution) and by which they maintain their existence (ecology)."[44] Leopold extends the conventional scientific axiom (since disputed) that "ontogeny repeats phylogeny" from the natural world to human experience: "the development of each individual repeats the evolutionary history of the race. This is true of *mental* as well as physical things."[45] This analogy posits a *correspondence* between landscape and culture. As Leopold himself asserts, "Ontogeny repeats phylogeny in society as well as the individual."[46] I would argue that this line of reasoning might be extended to the interlocking processes of forming individual and collective identities in indigenous cultures.

The journey recounted in *Books and Islands* takes on spiritual dimensions as Erdrich acquires, in situ, the language that engages a cultural legacy linking the Ojibwe people to this landscape. Recalling the names of medicinal plants, for example, Erdrich reflects on her own desire to acquire Ojibwemowin "to understand the prayers, and the *adisookaanug*, the sacred stories."[47] She frankly acknowledges the difficulty of learning this traditional tongue, particularly for an English-speaking adult like herself. In fact, she characterizes her own faltering first steps as an embarking on "what seems at times a quixotic enterprise—learning one of the toughest languages ever invented."[48] She reports, for example, that a single verb can take as many as six thousand forms. Moreover, Ojibwemowin proves particularly conducive to lexical innovation, adding greatly to its complexity: "The best speakers," she points out, "are the most inventive and come up with new words all the time."[49] Consequently, she relies on more fluent speakers, particularly Tobasonakwut—a native speaker of Ojibwe and respected tradition bearer who, McNab observes, is also "a sundancer" and a "fifth degree Medwewin medicine man."[50] "Learning Ojibwemowin," she concludes, "is a lifelong pursuit that might be described as living a religion."[51]

Acquiring the Ojibwe language, Erdrich discovers, entails tracing connections between indigenous culture and the natural world. Naming the figures on rock paintings, for example, some reproduced in the book by her drawings from the trip, invokes a subtext of narratives from oral tradition. McNab contends that "places[s] and the stories that come from them are an integral and inseparable part of nature."[52] "Songs," he points out, "are also stories. These were sung since ancient times on these islands."[53] Indeed, the Ojibwe traditionally associated *nagamonan* (songs) with specific places: "Songs belong to these islands," as Erdrich explains. "When Ojibwe people fast in these islands, the songs, even if lost for a time, always come back in dreams. The *nagamonan*. These very old songs are as old as the rock paintings."[54]

Though not all rock paintings can be interpreted with certainty today, Erdrich relishes the fact that we can readily recognize many figures. "I am standing before the rock wall of Painted Rock Island," she declares, "and trying to read it like a book. I don't know the language, though," enough to decode the "several spirit figures as well as diagrams of teachings" depicted there.[55] Yet she remains dedicated to contemplating their symbolic significance to the extent that cultural memory and imagination allow. "Whenever I offered tobacco," Erdrich confides, "I was for that moment fully there, fully thinking, willing to address the mystery."[56] In the process, she discerns how wilderness habitat supports populations of plants and animals on which the Ojibwe traditionally subsisted—and that still provide a basis for their spiritual relationship to the natural world. Images of these species among the pictographs represent a visual lexicon of indigenous knowledge and spiritual heritage.

While other tribes such as the Cree previously inhabited the region, the rock art found there is widely believed to be of exclusively Ojibwe origin, with much of it dating from the last four or five hundred years. Yet even the dating of the paintings remains speculative, largely based on oral tradition and inference. It is difficult to scientifically calibrate relative ages with any certainty by measuring erosion or the growth of lichen, for example, since the rate of such processes can vary greatly. Further complicating interpretation, images at a given site may have been painted concurrently, or be separated by centuries, since a place historically deemed sacred might inspire subsequent generations to add their own mark and thereby participate in an age-old tradition. Similar glyphs at separate sites painted at different times, moreover, may or may not have served the same function. Some figures undoubtedly correspond to mythical animals from traditional oral narratives. Yet stories for origins of the glyphs have morphed across generations, while others were simply lost. The accounts of Ojibwe elders preserved by even the most well-intentioned ethnographic transcripts collected in the late nineteenth century, such as those of Ernest Oberholtzer (1884–1977), are fragmentary at best.

Oberholtzer was an avid outdoorsman, pioneering conservationist, and self-taught ethnographer among the traditional Ojibwe in the region.[57] In fact, his relentless campaign to save the Quetico-Superior wilderness and to protect the vast watersheds of the Boundary Waters from hydroelectric dams ranks among the most audacious and unsung triumphs in the history of conservation in the Upper Midwest. Remarkably, his accomplishments ranged from transcribing stories from oral tradition longhand in Ojibwemowin to cofounding the Wilderness Society along with Sigurd Olson and Aldo Leopold. Biographer Joe Paddock captures the symbiotic relationship between Oberholtzer's devotion to the nascent environmental movement and his anthropological interest in recording and preserving Ojibwe culture and language: "He spoke in praise of native peoples and their traditional ways, which he believed should be studied as an extension of the wilderness itself. He also talked of Indians losing their culture and their story.... Ojibwe friends sometimes came to Ober to retrieve a story he had originally gotten from their grandparents."[58]

Widely admired by his Ojibwe neighbors, Oberholtzer was nicknamed Atisokan, an Ojibwemowin term for especially important stories from oral tradition such as the Nanabozho tales, believed to be particularly old, and told only at prescribed times of year. His anthropological fieldwork exemplified what has become known as salvage ethnography—reminiscent of turn-of-the-century photographer Edward Curtis's attempts to exhaustively document Native American peoples to preserve a glimpse of indigenous history. The work of both men was predicated on the conventional wisdom of the time that Indians were destined to disappear. Oberholtzer vividly remembered those who still practiced the old ways at the time: "They danced in the old way. They held the original Mide ceremonies, and they had their medicine men, and the *tchisiki-winini*...a magical being...mythological creatures that can talk to spirits."[59] In truth, the esoteric practices of the Mide society of shamans would never have been common knowledge, although there is evidence that he was permitted privileged access—virtually unprecedented for a European—and even participated in some of the most guarded ceremonies and rituals. During interviews in 1964, Oberholtzer described his undying desire to preserve in print as much of Ojibwe oral tradition as still existed while there was time since "the old people that could tell you these things are nearly all dead." Even at eighty years of age, Oberholtzer hoped to devote himself "to going around all through Ojibwe country trying to find the relics of things I once knew and record them."[60]

Oberholtzer's retreat on Mallard Island—with its extensive library centered on Native American cultures, above all the Ojibwe—is where Erdrich's journey ultimately leads. Little wonder that she would be drawn to such a legacy.

Textuality might be considered a central theme of *Books and Islands*. Rock paintings on the islands are likened to books, and the metaphor is signaled by the title itself. Indeed, writing provides the extended metaphor that frames the volume: islands with rock paintings that Erdrich likens to libraries and, by implication, the landscape itself inscribed with indigenous history and language. "The two have merged for me," she declares on the opening page. "Books, islands. Islands, books." The trope is sustained throughout her account: "These islands, which I'm longing to read, are books in themselves."[61]

Erdrich makes this connection explicitly as she describes her attempts to interpret glyphs by, in effect, *reading* them. Such graphic images arguably constitute a form of writing broadly defined. Among those remaining intact and *legible*, some have resisted the forces of erosion for a millennium or more. During the journey, Erdrich and companions actually encounter in the wild many of the animals depicted in the paintings. Identifying species as she attempts to interpret each glyph triggers a deeper process of memory and association. Her interpretations often incorporate knowledge of traditional Ojibwe culture, above all language, allowing her to more fully read these ancient traces in the landscape. Learning the Ojibwe names for figures depicted by specific glyphs, for example, including species of animals presumably endowed with spiritual significance, deepens the way she reads the rock paintings. Consequently, her desire to read and interpret the rock art she finds on islands in Lake of the Woods becomes entwined with her quest to acquire indigenous language. What clues might Ojibwemowin provide?

Ojibwe names for plants and animals reflect the history and lifeways of a people—the very point at which the human community meets and interacts with the natural world. Accordingly, a lexicon of traditional foodstuffs mirrors a specific *ecological* context as well as a cultural one. For Erdrich the relationship between the ecological and the cultural is mediated by indigenous language, perpetuating traditional Ojibwe values such as a deep reverence for nature—a characteristic of subsistence cultures expressed when offering tobacco, gathering medicinal herbs, tending beds of wild rice, or recognizing the threat of environmental degradation. Other contemporary writers concerned with cultural relationships to nature, such as

Barry Lopez, exhibit a sensibility that Slovic characterizes as "embodying a reverence for language itself."[62] Lopez himself believes that "what it means for writers is you have a responsibility that goes beyond yourself to care for the spiritual quality, the holy quality, the serious quality of language."[63]

Erdrich describes the level of veneration shown by those aspiring to speak Ojibwemowin: "Before attempting to speak this language, students petition [its] spirits with gifts of cloth, tobacco, and food."[64] "A spirit...belongs to each word," she explains; "the language is a deeply loved entity."[65] The practice of offering tobacco in this way cultivated a heightened awareness of the natural world, especially wildlife: "I found myself compelled to behave toward the world as if it contained sentient spiritual beings."[66] In fact, *asema* is frequently used throughout the journey, whether as a prayer at a sacred site or as a respectful gesture to another person. "Tobacco," she tells us, "is handed to anyone with whom you wish to speak in a serious manner. It is given for a story, or as an invitation to join someone in a teaching or writing project."[67] *Asema* is thereby woven into the social fabric of Ojibwe life and intimately linked to language itself. Just such a regard for Ojibwe language and tradition, after all, has led her to Lake of the Woods.

Erdrich further asserts that many of the remaining rock paintings are intelligible to the *modern* Ojibwe, evidence of the ongoing perpetuation of cultural identity. "Since the writing or drawings that those ancient people left still makes sense to people living in Lake of the Woods today, one must conclude that they weren't ancestors of the modern Ojibwe. They were and are the modern Ojibwe."[68] Traditional rituals are still being performed at some of these sacred sites, such as the practice of offering tobacco and prayer. Participating in such rituals at the sites of paintings, moreover, underscores the sacred dimensions of linguistic and cultural restoration. Expressing reverence in this way, and abiding by such customs, Erdrich seems to renew an ancient Ojibwe covenant for herself. Before embarking, she had fastidiously prepared offerings for "the spirits of the lake and spirits of the rock paintings we will visit. I gather gifts for the paintings—ribbon shirts, some red cloth, sage bundles."[69] She ultimately seeks to situate each painting in relation to the dynamic cultural processes at the time of its creation. Such ancient art seems to possess a power to engage the modern imagination. Moreover, the paintings represent a tracing of environmental history—or what Erdrich terms "spiritual geography." Taken together, they portray an entire historical ecosystem prior to the arrival of whites, one encompassing a multitude of species imbued with both biological and cultural significance.

Animal images sometimes suggest ceremonies related to hunting game. In subsistence cultures, after all, a successful hunt was of paramount—even existential—importance. Therefore, appeasing animistic deities, however they were conceived, was seen as essential to survival. Yet even when the depiction of an animal seems to be simply representational, it may have totemic dimensions, represent a clan, or signify a cycle of stories. In fact, some images may constitute a visual recording of mythic oral tradition, including stories no longer told. A single glyph might thereby serve as synecdoche for an entire animistic cosmology. The spiritual significance of certain figures can sometimes be discerned by comparing them to other artifacts, such as birch bark scrolls incised with similar images by the Mide society of Ojibwe healers and priests that Erdrich deems "perhaps the first books in North America."[70] While the sacred connotations may escape us today, such correspondences sometimes enable archaeologists to estimate the period when specific rock paintings were made.

In addition, an image may be layered with metaphorical or symbolic meaning, such as a canoe suggesting a journey, for instance, whether terrestrial or spiritual. Others seem more imaginary or even supernatural, associated with the spirit world evoked by dream or trance. In fact, several of Erdrich's early poems suggest shape-shifting between human and animal form, a transformation common in the traditional stories of many Native American peoples. For instance, "I Was Sleeping Where the Black Oaks Move," from Erdrich's first collection of poems, *Jacklight*, conjures the spirit of heron, linking a visceral sense of the bird's flight to the human dream world: "Sometimes now, we dream our way back to the heron dance. / Their long wings are bending the air / into circles through which they fall. / They rise again in shifting wheels."[71] Such poems seem to echo the metaphysical dimensions of the rock art.

Conceivably, the act of making such a painting constituted a prayer-like ritual, or even be linked to initiation, imbued with—or imparting—shamanistic powers. As Erdrich explains, lines "express relationships, usually between a human and a supernatural being," by suggesting forms of communication such as song, sound, or speech, while wavy lines "signify direct visionary information, talk from spirit to spirit."[72] Glyphs attributed with such esoteric functions remain among the least likely to be deciphered with any certainty, and accordingly the interpretation of such paintings remains speculative. Still, even the most "abstract" and enigmatic designs can enchant us with their aura of mystery. When Erdrich writes, "I've touched the rock paintings, and read a fragment of their stories," she is acknowledging the enigmatic character of many such images.[73] Consequently, it is essential to approach interpreting the figures with humility, much as Erdrich herself has done in *Books and Islands*, as she recognizes that the systems of symbolism may remain beyond our grasp—beyond time's horizon.

Yet Erdrich imagines that her own ancestors walked the very ground she now traverses, and she provides a brief account of her family's heritage, untangling its intercultural history and migration:

> The Ojibwe side of my family, who ended up with the surname Gourneau, roamed from Madeline Island in Lake Superior, along what is now the Canadian border, through Lake of the Woods and down to Red Lake, and then out onto the Great Plains and eventually the Turtle Mountains. Baupayakiingikwe, Striped Earth Woman, was one of those ancestors, as was Kwasenchiwin, Acts Like A Boy. Our family was of the Ajijauk or Crane dodem, and the Makwa or Bear dodem.[74]

The map of the region that appears as a frontispiece to the volume centers on this destination: Lake of the Woods straddling the Minnesota-Ontario border. The language was spoken here historically and, indeed, presumably "evolved" in the region, as Erdrich explains:

> Ojibwemowin is one of the few surviving languages that evolved to the present here in North America. For an American writer, it seems crucial to at least have a passing familiarity with the language, which is adapted to the land as no other language can possibly be. Its philosophy is bound up in the northern earth, lakes, rivers, forests, and plains. Its origins pertain to the animals and their particular habits.[75]

Knowledge of the species of plants and animals that inhabit this ecosystem has undoubtedly helped ensure the survival of the Ojibwe people for millennia. Heading from Red Lake to Lake of the Woods, Erdrich describes the region's extensive bogs as "traditionally the great Ojibwe pharmacy."[76] She recites the Ojibwe names for native plants found in these wetlands as well as a litany of their uses as foodstuffs—from highbush cranberries and wild currants to *miinan* (blueberries)—or medicines such as *makibug* (sumac) for dysentery and *winabojo nokomis winizisun* (painted cup) for rheumatism. Cattails, it turns out, hold a special place in Ojibwe medicine, as Erdrich explains: "Wiikenh, or sweet flag, is the star of the floating bog, though, a medicine with every possible use.… Wiikenh is the ultimate medicine."[77] She reports, for example, that "it is believed to strengthen the immune system and is mixed with water lily root to fight infection."[78] And its ritual uses include when "singers chew it to strengthen their voices at the drum."[79] As Tobasonakwut remarks, "Where there is wiikenh, there are Anishinaabe."[80] En route to the Boundary Waters, she characterizes the ethnobotany of wetlands along the highway where "countless" medicines grow in "the tamarack bog, an ecosystem so vibrantly rich that traditional Ojibwe teachers and healers still go out to fast there, to show their respect for the medicines and to learn from these plants."[81]

Traversing Lake of the Woods, Erdrich and her daughter encounter many of the animal species that thrive there, from *ashaageshiinhyag* (crayfish) to the otters that feast on them. In fact, by the time she prepares to return to Minneapolis, Erdrich realizes that she has witnessed every living thing depicted in the most impressive rock paintings: wild rice and deer, sturgeon and bear. She contemplates how the *bizhiw* (lynx) became the totem of her infant daughter's ancestors, whom the French explorers referred to as "People of the Cat."[82] Her daughter was named for a spirit woman from Ojibwe oral tradition, but that name turns out to involve kinship as well as a mythic dimension: "The baby in the car seat behind me is named for that spirit woman, Nenaa'ikiihikok. Her grandmother on her father's side had this name, and was called Kiizhikok, Sky Woman, or Kiizhik; Sky, for short."[83] Reflecting on her child's paternal grandmother and namesake in this way raises the thorny question of perpetuation of Ojibwe language and tradition across generations in the face of cultural contact and assimilation. She realizes that "the only image I have of her [grandmother] is from the back row of a boarding school class picture.… This Nenaa'ikiihikok is spoken of in Ojibwemowin as Nenaa'ikiihikokiban. The 'iban' at the end puts her in the past, in the spirit world, where I imagine she is still dancing in her jingle dress. She was a well-known expert jingle-dress dancer, and even came to the Turtle Mountains to powwow."[84]

Throughout their journey on the lake, Erdrich repeatedly marvels at how various animals in the wild—the normally timid *waboose* (rabbit) and the reclusive *maang* (loon), even the *nameh* (sturgeon), as well as otter, moose, eagles, and pelicans[85]—appear to respond to her infant daughter with what can only be called a kind of curiosity. Erdrich muses that the child has somehow inherited a spiritual relationship to this place—perhaps a distinctly Ojibwe one. She tenderly portrays meeting her infant's gaze in a strikingly lyrical and metaphysical way in the poem "New Mother," published the same year as *Books and Islands*:

> When I look into her eyes, I see below
> the surface of things
> into the water of the other surface
> through the layers of that surface
> to the original fire.[86]

Each creature evokes a cultural legacy. The Ojibwemowin word for turtle, *mikinaak*, is connected with the Ojibwe calendar of thirteen moons (there are thirteen plates on the turtle's back), which in turn leads Erdrich to reflect on Ojibwe notions of numbers. "Mathematics wasn't abstract," she tells us. "It was intimate. Dividing and multiplying and factoring were concerns of the body."[87] She speculates that attention to cycles of fertility led women to make calculations that form the basis of indigenous mathematics. Consequently, she explains, the turtle found in rock paintings is still associated with women among the Ojibwe. To call the wolverine by its Ojibwe name—*kwiingwa'aage*—summons its "almost supernatural reputation." The Ojibwe creation story portrays wolverine as a celestial creature with unequalled prowess as a hunter—an animal spirit to be revered. "A spirit of dark strength and cleverness.... There is an animal who follows you just out of sight. It is deathless, lonely, and somewhat strange in his contempt for human intelligence. He easily outwits the smartest hunters. When the creator passed near the Earth in the form of a tailed light, that was the *kwiingwa'aage.*"[88] *Kwiingwa'aage*. Speaking this single word invokes a story of creation, signaling reverence for a totemic figure and reflecting the Ojibwe worldview rooted in relationship to nature.

One rock painting that especially captivates Erdrich is of the wild rice spirit, *manoominikeeshi* to the Ojibwe. She sketches this stylized figure that faithfully represents the delicate roots of "a *spiritualized* wild rice plant."[89] Historically, such wild edibles helped ensure the survival of subsistence cultures in a particular place over generations. Prior to European contact, after all, Indians thrived in the Americas for millennia through what often amounted to sustainable use of the natural "resources," including wildlife, reflecting a sacred relationship based on an ethic of reciprocity. Failing to do so could cause a crash in game populations and necessitate relying on other foodstuffs, perhaps even relocation. Similarly, gathering edible plants in the wild entails an essential element of stewardship, taking care not to overharvest, as well as rituals associated with propagation and cultivation.

Current environmental changes threaten the integrity of traditional cultures. Erdrich laments that rising water can jeopardize nearby wild rice beds still harvested by Anishinaabe families, and asserts that "the water level is a political as well as a natural process,"[90] a process that McNab dates to at least twenty-five years before her travels there.[91] Apparently, such fluctuations are sometimes precipitated by the opening of floodgates at dams and mills on watersheds upstream. "This year on Lake of the Woods, the rice looks dismal," she reports, "because the high waters have invaded.... The water ruined thousands of acres of wild rice beds."[92] She reports that some of the ancient paintings are already underwater, and that rising waters threaten additional rock art as well: "When the pictures were painted, the lake was a full nine feet lower, and as it is nearly four feet higher this year than usual, some paintings of course are submerged."[93] The local Ojibwe blame the provincial government's management policies for this degradation of natural and cultural resources. The vulnerability of this ancient art shows how environmental change can disrupt indigenous relationships to nature.

The extinction of species can have especially dire consequences for Native peoples. Consider the massive *nameh* (sturgeon) depicted as a totemic creature in the rock paintings. The ancient fish dates back at least 265 million years and is a potent symbol of indigenous resilience and survival. Historically, the biological connection between the Ojibwe people and this landscape was paramount. Indeed, Erdrich tells us, "At one time, everyone who lived near the lake was essentially made of the lake. As the people lived off fish, animals, the lake's water and water plants used for medicine, they were literally cell by cell composed of the lake and the lake's islands."[94] This physical relationship to place, of course, complements a spiritual one mediated by the Ojibwe language. As Leopold has observed, a profound "value" of wildlife is its power to *remind* us, symbolically as well as literally, "of our dependency on the soil-plant-animal-man food chain, and of the fundamental organization of the biota."[95] And while such connections to natural ecosystems can admittedly be obscured by modern commerce and manufacturing, such an essential relationship is surely an archetypal element of human experience and conceivably a cornerstone of culture.

Once abundant throughout the region, Erdrich recounts, sturgeon have been largely extirpated by overharvesting by commercial fisheries.

> The sturgeon up here on Lake of the Woods were the buffalo of the Ojibwe. Greed and overfishing by non-Indians caused their population to crash around the turn of the nineteenth century, when, along with other Great Lakes, Lake of the Woods became one of the world's principal suppliers of caviar. The sturgeon were indiscriminately taken by the non-Indian fisheries for their roe, much as buffalo hunters took only the buffalo tongues. They were stacked like cordwood all along the lake and often left to rot. An agonizing sight for the Ojibwe, who revere the sturgeon and know its secrets.[96]

This tragically apt comparison to the buffalo underscores the environmental stakes. By the mid-nineteenth century, as William Rossi describes, John James Audubon and Henry David Thoreau were already lamenting the decline of species: "Habitat was being irreparably altered and destroyed, unregulated hunting, fishing, and trapping was decimating populations of moose, wolf, deer, bear, and migrating fish such as salmon, shad, herring, and alewife that Thoreau could remember from boyhood."[97] In *The Ecological Indian*, Shepard Krech recounts the brutal demise of the bison as emblematic of historical forces that threatened Indian peoples:

> Today there is almost universal agreement that the near extinction of the bison is one of the grimmest narratives in the history of wildlife. At the end, the decline west of the Mississippi was rapid and extravagant to the extreme, because hunters only stripped the robe from each buffalo they killed, leaving carcasses to bloat in the sun. Indians uniformly mourned the passing of the bison. But not whites; some spoke of buffalo as noble and majestic, but others were unconcerned, and others actively promoted extermination as inevitable in the westward course of empire and as a progressive impetus in bringing civilization to native people.[98]

The consequences for those tribes most reliant on bison were especially dire, as Leopold underscored in his essay "Wildlife in American Culture": "The plains Indian not only ate buffalo, but buffalo largely determined [their] architecture, dress, language, arts, and religion."[99] Erdrich herself poignantly recounts the indiscriminate slaughter and devastation of the herds in her poem "The Red Sleep of Beasts."

We lived headlong, taking what we could
But left no scraps behind, not like the other
Hide hunters, hidden on a rise,
Their long-eyes brought herds one by one
To earth. They took but tongue, and you could walk
For miles across the strange hulks.

...the old ones knew
They would not come again to the low hills.[100]

Yet like the buffalo, the sturgeon has so far escaped extinction, aided by conservation measures such as hatcheries established by Canadian agencies as well as tribal communities. However, the excessive stocking of hatchery-raised fish has had a detrimental impact on populations of wild native species. Stock bred from an indiscriminate mixture of populations, for example, by drawing from different watersheds, disrupts natural breeding cycles, commingles distinctive gene pools, and thereby compromises the genetic integrity and diversity of a species. Perhaps most egregiously, architects of such fishery policies too often assume that hatchery-raised fish can "mitigate" the ravages of wholesale habitat destruction. Such concerns are hardly new. In the celebrated *Sand County Almanac*, first published in 1949, Leopold ponders a similar paradox: "Consider...a trout raised in a hatchery and newly liberated in an over-fished stream. The stream is no longer capable of natural trout production. Pollution has fouled its waters, or deforestation and trampling have warmed or silted them. No one would claim that this trout has the same value as a wholly wild one." Precisely because this hatchery trout has been "artificialized," it lacks the symbolic cachet of wilderness legacy. To make matters worse, Leopold goes on, managers can rapidly compound the damage by exterminating predators such as "all herons and terns visiting the hatchery...and all mergansers and otters inhabiting the stream."[101]

Nonetheless, when a native population has been overharvested to the point of threatening the existence of a species, intensive breeding in captivity can indeed provide a stay against extinction. Such proactive management strategies can also counter local extinctions caused by fragmented habitat. Even critics concede the value of relying on hatcheries in certain circumstances: "They can and do serve as genetic reservoirs, saving species from extinction until habitat is repaired." Indeed, "without hatcheries to hold rare stock all but lost in the wild, there would be no self-sustaining populations of a species including several varieties of salmon and trout as well as the pallid sturgeon."[102]

Such efforts to sustain sturgeon at Lake of the Woods offer a poignant reminder of its symbolic significance to the Ojibwe. Some rock paintings of sturgeon suggest a distinctly spiritual relationship, depicting the great fish beside a triangle Erdrich describes as "a divining tent, a place where Ojibwe people have always gone to learn the wishes of the spirits and to gain comfort from their teachings."[103] In fact, the binder for these paints was derived from sturgeon by rendering the cartilaginous skeleton and mixing it with pulverized stone to produce a sacred shade of ochre. A remarkably durable paint, or *atisikan*, concocted from what she terms "mysterious ingredients" gathered locally, has helped ensure that ancient rock art endures. When Erdrich finally reached the place known to the Ojibwe as "Eternal Sands" (*Kaawiikwethawangag*), she discovered for herself the vivid red roots of beach plants that mixed with sturgeon oil had enabled rock paintings to last for centuries. Her pilgrimage had led to a place where such roots still grow.

As Nelson has argued, among "the fundamental tenets of the creative vision informing American Indian" literature—including the masterworks of several contemporary novelists—"is the proposition that place, understood as a living physical landscape invested with the same type and degree of spirit as humanity, has the power to shape the identities of the People, individually and collectively." Erdrich's reflections on her journey to the traditional Ojibwe homeland at Lake of the Woods underscore Nelson's premise that "place brings the individual into identity also with life of the People...[and that] collective traditions have evolved out of...a similar process."[104] In *Books and Islands*, Erdrich expressly reflects on this multifaceted process of establishing an indigenous identity in relation to place. As Lischke concludes, "Erdrich tells stories, indeed her life story, to find her own sense of self, her place, and her spirituality."[105] The ecosystems that she explores in *Books and Islands*, habitat for many kinds of wildlife that have long served as sources of subsistence, still animate Ojibwe culture: incorporated into oral tradition, traced in the rock paintings, and thereby inscribed into the landscape itself.

NOTES

1. While ecocultural and biocultural perspectives provide a theoretical framework for ethnographic and sociological studies in a wide variety of contexts, the term is intended here in a more general thematic and descriptive sense.
2. Erdrich, *Books and Islands*, 14.
3. Treuer, *Native American Fiction*, 66.
4. Ibid., 62.
5. Ibid., 64.
6. Ibid., 49.
7. Ibid., 56.
8. Erdrich, *Books and Islands*, 87.
9. Ibid., 5.
10. Lischke, "Reflections on Metis Connections," 40.
11. John Nichols, qtd. in Vizenor, *Summer in the Spring*, 18.
12. Erdrich, *Books and Islands*, 81.
13. Lischke, "Reflections on Metis Connections," 40.
14. Erdrich, qtd. in Lischke, "Reflections on Metis Connections," 43.
15. Erdrich, *Books and Islands*, 81.
16. Ibid., 84.
17. Ibid., 81.
18. Ibid., 85; emphasis mine.
19. Ibid., 125.
20. Ibid., 85.
21. Leopold, *Sand County Almanac*, 188.
22. Ibid.
23. McNab, "Significance of Place," 11.
24. Buell, *Environmental Imagination*, 267; emphasis mine.
25. Glotfelty and Fromm, *Ecocriticism Reader*, xviii–xxi.
26. Buell, *Environmental Imagination*, 7.
27. Slovic, "Nature Writing," 352; emphasis mine.
28. Rossi, introduction to "*Wild Apples*," xiv–xv.
29. Ibid., xv.
30. Ibid., xiii.
31. Slovic, "Nature Writing," 364.
32. Buell, *Environmental Imagination*, 20.
33. McNab, "Significance of Place," 207; emphasis mine.
34. Nelson, conclusion to *Place and Vision*.
35. Bunge, "Mystical Midwest."
36. Ibid., 113–14.
37. Lischke, "Reflections on Metis Connections," 50.
38. Ibid.
39. McNab, "Significance of Place," 218.
40. Nelson, introduction to *Place and Vision*.
41. Lischke, "Reflections on Metis Connections," 44.
42. Nelson, introduction to *Place and Vision*.
43. Lischke, "Reflections on Metis Connections," 40.

44. Leopold, *Sand County Almanac*, 173.
45. Ibid.; emphasis mine.
46. Ibid., 178.
47. Erdrich, *Books and Islands*, 81.
48. Ibid., 82.
49. Ibid., 83.
50. McNab, "Significance of Place," 208–9.
51. Erdrich, *Books and Islands*, 87.
52. McNab, "Significance of Place," 205.
53. Ibid., 210.
54. Erdrich, *Books and Islands*, 38.
55. Ibid., 51.
56. Ibid., 16.
57. A detailed account of his wide-ranging accomplishments can be found in *Keeper of the Wild: The Life of Ernest Oberholtzer*, by Joe Paddock.
58. Ibid., 276.
59. Ibid., 277.
60. Ibid.
61. Erdrich, *Books and Islands*, 3.
62. Slovic, *Seeking Awareness*, 148.
63. Lopez, qtd. in ibid., 148–49.
64. Erdrich, *Books and Islands*, 87.
65. Ibid.
66. Ibid.
67. Ibid., 14.
68. Ibid., 6.
69. Ibid., 7.
70. Ibid., 5.
71. Erdrich, *Original Fire*, 15.
72. Erdrich, *Books and Islands*, 56.
73. Ibid., 136.
74. Ibid., 80.
75. Ibid., 85.
76. Ibid., 24.
77. Ibid., 59.
78. Ibid., 64.
79. Ibid.
80. Ibid., 59.
81. Ibid., 25.
82. Ibid., 32.
83. Ibid., 12.
84. Ibid., 12–13.
85. Ibid., 58, 75.
86. Erdrich, *Original Fire*, 133.
87. Erdrich, *Books and Islands*, 62.
88. Ibid., 40–41.
89. Ibid., 51; emphasis mine.

90. Ibid., 54.
91. McNab, "Significance of Place," 214.
92. Erdrich, *Books and Islands*, 51.
93. Ibid., 52.
94. Erdrich, *Books and Islands*, 34.
95. Leopold, *Sand County Almanac*, 178.
96. Erdrich, *Books and Islands*, 75–76.
97. Rossi, introduction to "*Wild Apples*," viii.
98. Krech, *Ecological Indian*, 126–27.
99. Leopold, *Sand County Almanac*, 177.
100. Erdrich, *Original Fire*, 31.
101. Leopold, *Sand County Almanac*, 169.
102. Williams, "Something's Fishy," 273.
103. Erdrich, *Books and Islands*, 77.
104. Nelson, conclusion to *Place and Vision*.
105. Lischke, "Reflections on Metis Connections," 41.

8

Giving Voice to History

Diane Glancy (b. 1941)

Because when grief set out on the road
the people went and arrived
and united new land and water
to sow their word once again.

—Pablo Neruda, "The Word"

Beginning with her first published volume of prose, *Traveling On*, in the early 1980s, Cherokee poet and novelist Diane Glancy has emerged as a major Native American voice, a remarkably prolific and highly versatile writer who has published twenty volumes of fiction, as many collections of poetry, and nearly a dozen plays, as well as several books of essays. In addition to her impressive range in terms of genre, she has experimented boldly with form, using multiple narrators in historical fiction, for example, as well as intertextuality and hybridity when incorporating source documents. This chapter examines a pair of linked novels from an ecocritical perspective to illuminate the role of nature and place in the formation and perpetuation of indigenous language and cultural identity. *Pushing the Bear: A Novel of the Trail of Tears* (1996) is set immediately preceding and during the forced march from September 1838 to February 1839. The sequel, *Pushing the Bear: After the Trail of Tears* (2009), takes place following relocation from southeastern states, spanning Alabama, Georgia, North Carolina, and Tennessee, as the tribe seeks to preserve its beliefs and reestablish traditional lifeways in the unfamiliar landscape of today's Oklahoma, then known as Indian Territory. What would become of their customs, language, and traditions, these novels ask, having left their ancestral homelands, especially farms in the woodlands of North Carolina?

Glancy has a personal stake in this history, describing her own Indian heritage geographically in relation to relocation: "My father was an undocumented Cherokee because his father was not in Indian Territory at the time of the Dawes Rolls."[1] Her forebears were presumably among those who eluded capture and escaped relocation by retreating to increasingly remote parts of the Southeast's unsettled woodlands. Glancy herself is a genuine denizen of the Midwest, with an unmistakable attunement to place. Born in Kansas City, Missouri, she attended both the University of Missouri and University of Iowa before teaching at Macalester College in Minnesota, living subsequently in Kansas and Oklahoma. In fact, place figures directly in her creative process, especially to establish narrative voices. She explains that while preparing to write each of her four historical novels, she made journeys "in search of voice."[2] In the case of *Pushing the Bear* she took several trips to northeastern Oklahoma that resembled pilgrimages: "I drove to the place where the history took place.... I think that it was in passing over the land that I began to hear the voices of the people in my imagination," she explains. "I am interested in giving voice to history."[3]

Both volumes of The Trail of Tears sequence grapple with the loss of verdant homelands and the enormous loss of life when an estimated eleven thousand to thirteen thousand Cherokee were made to march nine hundred miles on foot in the dead of winter. The grueling journey would stretch across Kentucky, Illinois, Missouri, and Arkansas before ending at Indian Territory. During months of snow squalls and rainstorms, as well as treacherous river crossings at the Ohio and Mississippi in midwinter, one in four perished. Glancy's account does not spare us the raw misery, grisly hardships, and outright atrocities of the march, but portrays equally the psychological trauma. In fundamental ways, relocation was a profound affront to the Cherokees' sense of cultural identity. Place-based cultural memory was jeopardized as soon as the column of refugees—initially ten miles long—were marched westward toward in Tennessee in October 1838. Their ancient relationship to familiar forests was abruptly severed, utterly disrupting age-old lifeways. The Cherokee were understandably apprehensive that rumors about land farther west might prove true: what if this new territory lacked the biological bounty the tribe had always been able to rely on in the past? Those who lived to see Indian Territory were stripped forever of many of the natural resources that had been essential for their survival. Indeed, their identification with the land of their ancestors—which is to say an amalgam of nature, place, and narrative—was at times so visceral as to seem corporeal.

As the first of the two *Pushing the Bear* novels commences, the Cherokee in North Carolina are caught up in the ugly flux immediately preceding removal, exacerbated by the cruelty of callous soldiers. In a whirlwind of change, white settlers wasted no time in staking claim to farms and giving English names to place. There was a widespread sense of betrayal over Andrew Jackson's Indian Removal Act, a blatant land grab that Congress had authorized by a single vote. Moreover, many were deeply bitter at having been sold out by other bands in Georgia. A renegade faction signed the Treaty of New Echota in 1835, ceding the tribe's lands in the east, including North Carolina and Tennessee, without the knowledge or consent of the Cherokee still living there. "Thirty-five men signed away land," Glancy tells us, "that belonged to seventeen thousand Cherokee."[4] Being brutally uprooted from woodland farms also meant abandoning landscapes where they had learned to listen to the voices of ancestors for solace and counsel. They could only hope that such guiding spirits would somehow accompany them westward. On the trail to Indian Territory, the spectral voices of ancestors were perceived not as remembrance, but actual presences.

Native Americans rounded up for removal in 1838 were far from being a homogeneous population, especially since disparate clans were relocated from several different states in quick succession. *Pushing the Bear* characterizes the throng as "a mix of diverse peoples. Agreeing on little. Our seven clans divided between three white peace clans and three red war clans.... Small farmers, many of us illiterate. Plantation owners. Slaves. Half-breeds. Whites who'd intermarried. Conjurers. Christians."[5] Like a number of major Native American works employing multiple narrators, such as N. Scott Momaday's groundbreaking *The Way to Rainy Mountain* or many of the celebrated novels of Louise Erdrich, Glancy's account of the Trail of Tears is told in a striking variety of voices. In truth, she has repeatedly experimented with alternating narrators in other works of historical fiction. *Stone Heart: A Novel of Sacajawea*, for example, juxtaposes an imagined account by the young Shoshone translator for the Corps of Discovery expedition with excerpts drawn from the historical journals of Captains Meriwether Lewis and William Clark. The first *Pushing the Bear* volume has an especially complex mix of perspectives, portraying a moment of profound cultural upheaval. By Glancy's own count, there are nineteen historical characters, in addition to at least twenty fictional ones. Nonetheless, she contends that since "the characters are speaking as a tribe...readers should not try to track the individuals as much as listen to the voices washing over them."[6] She describes this narrative strategy as a "communal first-person" point of view—an indigenous worldview she characterizes as "one-as-all, as opposed to Western culture's emphasis on the individual."[7] In fact, one of the novel's primary themes concerns the implications of acculturation, as some of the Cherokee succumb to European individualism, greed, and self-interest at the expense of the well-being of the People as a whole.

In the first of the two novels, Maritole emerges as one of the principal narrators enduring the Trail of Tears with her family—including her mother and an infant child who both die early in the journey. "Nothing would be like the land we came from," she declares while crossing Tennessee bound for Indian Territory. "Nothing would be like the ceremonies we had there."[8] Indeed, the arduous march to Indian Territory is fraught with uncertainty: with every day's walk the landscape changes before their eyes. As the journey progresses, they try to imagine the topography that awaits them. They hope Indian Territory will possess woodlands akin to those in North Carolina. As they continue across Tennessee, and the terrain alternates between woodland to prairie, many fear their new home will turn out to be arid and barren. After crossing grasslands farther east, however, they finally begin seeing more trees again.

Maritole's husband, Knobowtee, also begins to ponder the possible consequences of leaving the familiar landscapes of their homeland, a place replete with voices of nature that they had heeded: "The woods were growing thinner. Soon we'd be in an open clearing. What would it be without the hills of North Carolina? What would it be without the voices of the trees to help us walk?"[9] The forest that surrounded their farmsteads in North Carolina had actually foretold their removal: "Of all the trees who spoke," a female elder named Quaty Lewis tells us, "it was the pine who said we would be going, covering its needles with its hands."[10] Another elder, Anna Sco-so-tah, recounts the voice of the land she had heard as the soldiers dragged her away from home: "The birds called to me across the trees. Even the water spoke."[11] These passages reflect what environmental historian Timothy Silver terms the Cherokee "belief in the spirituality and human volition of plants and animals whom they regarded as their spiritual kin."[12] All are given voice and agency in Glancy's narrative.

Given their unique geological history, Cherokee ancestral homelands in the Appalachians were in fact graced with exceptional levels of biodiversity—including a host of ancient species, many of which still flourish in the region today. Relocation not only disrupted the tribe's place-based subsistence practices, but subsequently threatened to destabilize age-old rituals, including medicinal traditions. A holy man in the first novel, for example, laments that five species of trees whose bark has curative properties, including apple, dogwood, small acorn oak, wild cherry, and willow, were being left behind and potentially lost to the tribe forever. "What if those trees don't grow outside Cherokee land?" some wonder.[13] Renewing their search for familiar plants in Indian Territory, they are relieved to see at least a degree of ecological continuity: some trees, for example, are among those they recognize from the East, including redbud, black cherry, poplar, shagbark hickory, sweet gum, loblolly pine. Conjurers begin to gather medicinal bark again from local acorn oak, bitter apple, cherry, and white willow trees. This is the crux of ethnobotany: indigenous knowledge of species that grow in specific ecosystems, including the best spots to search for them and the most promising times to harvest. Leaving behind traditional homelands threatens to upend highly localized customs such as the therapeutic use of botanicals, while also jeopardizing the stability of cultural memory generally. Forfeited in the process is an intimate relationship with the places that had grounded Cherokee traditions for millennia. Consequently, what resettlement ultimately meant for those who survived, Glancy contends, was "a shift from everything they had known. They would have to rebuild in the new territory with what they could remember, but it too had shifted."[14] What was at stake—beyond the loss of their homes and most material possessions—was their very livelihood and, indeed, tribal identity itself.

In Glancy's estimation, Native American *language* is intricately linked to the places where it historically evolved, connections much like those that Louise Erdrich explores in *Books and Islands in Ojibwe Country*. What will become of the Cherokee language, they wonder, after leaving North Carolina, enduring the Trail of Tears, and arriving at the dramatically different landscapes of Indian Territory? Maritole ponders the fate of their language after being forced to abandon ancestral homelands: "Our words came out of the land. Would our words leave their place? Did they walk, too? Would we have language when we got to Fort Gibson in Indian Territory? Would we find our words hadn't migrated? If the language didn't cross the land, it would sink back inside the trees and rocks. It would be buried in the land. It would go back to where it came from."[15]

Glancy's narrative gives us a visceral sense of the Cherokees' emotional turmoil as well as a glimpse of how they tried to make sense of this experience. Traversing the Trail of Tears month after month, they continue to grapple with the profound disorientation and vertigo of dispossession. They contemplate the role of stories, for instance, likening them to baskets that "were woven from the land" like the Cherokee themselves.[16] "How much more like a basket were our stories? They held our fear and hurt and resentment and anger. They gave a place to order our disorder, a direction for our directionlessness."[17] Even after being removed, they continue to identify deeply with those ancestral homelands, regarding them as the center of the world. Loss of those landscapes stripped them of the very foundations of cultural identity, "had opened us up to the old disorder, and we would have to build the world again—and that took so much time and energy. They were unweaving the weaving that we had taken generations to complete."[18] Ultimately, they would need to begin filling this emptiness by weaving their culture all over again in relation to the unfamiliar landscapes of Indian Territory. Yet the outcome of such a transition process was unpredictable: "How long before the old territory left them?" Glancy asks. "How long before their old nature turned to new nature?"[19]

In the *Pushing the Bear* sequence, Glancy depicts Cherokee culture in its geographical context, particularly in terms of environmental history. She recognizes how the lifeways and traditions of the region's indigenous peoples had been shaped by these geological forces in deep time and then reshaped through millennia of ecological change. The first of the two novels, she explains, is meant to perpetuate the cultural memory that effectively stretches toward time immemorial: "a period extending to antiquity, beyond the record and memory of man."[20] Anthropologists have established that Indian peoples have inhabited the eastern woodlands for at least the last twelve thousand years—starting with the Paleolithic predecessors of the Cherokee, whose arrival in the region presumably followed the retreat of the Laurentide Ice Sheet, which began an estimated eighteen thousand years ago. The Appalachian Mountain range had been spared this final wave of glaciation; so its extraordinary biodiversity is due in part to having functioned as a refugia for many species that have continued to coevolve in complex and place-specific ways.

As the climate continued to warm, northern trees associated with tundra and boreal forests gradually gave way to the species found in the region today: these were the foundation of the verdant ecosystems that in turn sustained the Cherokee, shaping their language and culture. As Silver concludes: "The commingling of northern and southern animals in the highest

Appalachians suggests that these are not simply northern habitats that have somehow been preserved intact. Instead, what has emerged on these elevated slopes is a unique, southern, high-mountain forest."[21] Conifers such as fir and spruce were in turn supplanted by hardwood stands composed of beech, birch, buckeye, and, at lower elevations, chestnut, hickory, maple, poplar, and oak. Such dramatic shifts in forest composition—from boreal and northern hardwood to woodlands associated with temperate zones—inspired theories of plant succession proposed at the beginning of the twentieth century by pioneering ecologists such as Frederic Clements. However, the long-term environmental history of the region is considerably more complex than such tidy models might suggest. More nuanced thinking about succession takes into account periodic disruptions to any existing order—whether naturally occurring events such as prolonged drought, or human alteration such as clear-cutting. Accordingly, ecosystems in perpetual transition are by no means the singular or fixed entities suggested by theories that posit specific "climax" communities as the ultimate result of succession. They are more accurately characterized by highly dynamic biological processes that continuously reshape them.

Today's Cherokee people—like the preceding generations subjected to relocation on the Trail of Tears—are descendants of the Pisgah culture of Woodland Indians who cultivated crops in conjunction with other customary subsistence activities, including hunting, fishing, and gathering. Glancy recognizes how the cultures of Native American nations were historically shaped by—and in turn shaped—the ecology of specific landscapes. Preceding European settlement, in fact, indigenous peoples in the southeastern region of the Atlantic Seaboard "had altered the landscape to suit their needs, adapting techniques and technology known across the continent to the peculiarities of local climates and topography."[22] They had pioneered and refined methods for cultivating corn in the region, adapting agricultural practices to variations in terrain and elevation. Indeed, such adaptability and versatility had enabled the Cherokee people to thrive, reaching a population of at least thirty thousand by the time whites arrived.

Successful subsistence practices entailed knowledge of natural cycles that was intricate and exacting as well as place-specific: an attunement to changing seasons coupled with a keen appreciation for microclimates created by even subtle variations in a landscape of wooded hillsides and hollows. Forest composition reflected topography, such as where coves, ravines, and valleys sheltered stands of trees, producing a variety of habitats, each with its own complement of game to hunt and edible plants to gather. The sequence of subsistence activities was adjusted continually to stay in synch with seasonal as well as landscape change. Over the course of many generations, the Cherokee had learned to coordinate their hunting, gathering, and planting with the natural cycles of the forest ecosystem as a whole, such as when journeying to intercept an anticipated annual migration. Near the Atlantic coast, for example, spawning runs of freshwater and anadromous species such as alewives, herring, shad, and sturgeon in the rivers each spring signaled the start of an especially bountiful, albeit brief season. In addition, a variety of fish and shellfish were periodically abundant along the shore. Summer yielded a profusion of small game such as rabbits and squirrels, as well as wild edible plants including a variety of berries. The Cherokee returned to spots that had proven especially productive in the past. In autumn, nuts such as acorns, chestnuts, and hickory were gathered. These same trees also provided mast for a wide variety of foraging flocks, including passenger pigeons and wild turkeys as well as multitudes of migrating waterfowl. Deer were hunted intensively from the beginning of the autumn rut, making venison a winter staple.

Until the beginning of the nineteenth century, the Cherokee were the beneficiaries of the natural abundance of their ancestral homelands, woodlands that had supported a variety of age-old subsistence practices ultimately augmented by agriculture. "It was not a static world of ecological perfection or continuous bounty," as Silver concludes, "but, rather, a world in which people understood the land and its ways, where natives enjoyed easy abundance when they could but had no qualms about taking steps to ensure sustenance when nature turned harsh and unpredictable."[23] Since the eastern woodlands had been inhabited for millennia, the Cherokee had inherited fields and often used fire to clear brush and revitalize the soil with ash rich in nitrogen. Similarly, crops such as corn were planted at a particular time of year to coincide with the growth of indigenous species, reflecting an intimate knowledge of surrounding landscapes. As Glancy's characters explain, "We plant our outer fields when the wild fruit is ripe because the birds are busy eating it"; conversely, "the underbrush was burned off the fields near our farm so the deer would come and eat the corn and we could harvest their meat."[24] Once soil fertility had been exhausted at one site, abandoned fields would begin a centuries-long process of forest succession toward a diverse forest with native species of trees, typically beginning with pines, followed by a mixture of maples and dogwood, then ultimately chestnut, hickory, and oak. Even recently cultivated plots soon supported hunting again by providing excellent forage for major game species.

Cherokee agriculture had evolved over centuries prior to colonization: from gathering plants native to the region to cultivating crops that had originated elsewhere, such as eastern flint corn and at least three types of beans. Primitive gardening in the region dates back at least five thousand years and includes the domesticated descendants of edible wild plants such as gourds, squash, and sunflowers. Corn arrived in the foothills of the Appalachians and along its watersheds around eighteen hundred years ago. Over several centuries it became an intensively cultivated staple, supplementing a diet of nutritious forest foods such as acorns and hickory nuts. By one thousand years ago, the cultivation of crops, including corn, had become more entrenched, extensive, and elaborate—offsetting shortages from less predictable subsistence activities such as hunting and gathering, both subject to fluctuations in wildlife populations and the vagaries of weather.

The ancestral homelands the Cherokee were forced to abandon in the 1830s in the wooded uplands of North Carolina had historically boasted a fabulous natural bounty. With the spread of white settlement, however, wildlife populations steadily declined, making it increasingly common for Native Americans in the region—including those who would ultimately be relocated in Indian Territory—to procure meat by raising livestock of their own, so that they came to rely on slabs of cured ham instead of fresh venison or squirrel. Before livestock began to replace wildlife as a food source, cultivated crops such as pumpkins, beans, and squash also augmented diminishing subsistence resources such as game animals. By necessity the Cherokee had begun to forge a hybrid lifestyle, one that incorporated various European agricultural practices. Many homesteads abandoned during removal included livestock such as cattle, hogs, and horses, as well as harnesses, plows, and other farming implements; fields were cleared and fenced for crops, along with orchards of pears or plums, but above all, apples and peaches. Moreover, the Cherokee had also experimented with organized governance, including a constitution, while even achieving literacy, enabled by Sequoyah's visionary syllabary, contributing to their reputation at the time as a "civilized tribe." "We farmed to prove to them we were civilized," Glancy writes, "then they took our farms. We emulated the white man. Established a capital. Took power from the women. Made a two-party government."[25] Yet acquiescing to the mounting pressures of assimilation could only forestall a fate that came to seem inevitable for the Cherokee, as for so many other tribes: dispossession. Moreover, along the Trail of Tears, and then again when resettled in Indian Territory, they would be reduced to a demoralizing diet of government-issued commodity rations, including staples such as flour and sugar as well as cured meats such as bacon and salt pork, instead of familiar foodstuffs such as venison, chestnuts, and above all corn.

In truth, the southeastern coastal plains had seen considerable landscape change since colonists began settling the area in the sixteenth century. Before overhunting by European settlers, game was abundant throughout the region, including numerous herds of buffalo and elk, thriving populations of major predators such as bobcats, panthers, and wolves, and omnivores such as black bears. However, in the wake of rampant trapping and overhunting that accompanied settlement by whites, the region's indigenous peoples resorted to increasingly limited diets featuring staples such as cornbread and squirrel meat. Wherever the fur trade took hold, overharvesting of pelts quickly ensued, leading to widespread eradication of game. By the beginning of the nineteenth century, wildlife had already suffered precipitous declines:

> The rivers and creeks would hold fewer sturgeon and ocean-going fish. Freshwater fish, too, might be less abundant.... Fewer turkeys and waterfowl...dwindling flocks of Carolina parakeets and passenger pigeons. Buffaloes would be gone.... Beavers, muskrats, otters, and minks also would have diminished.... Wolves, panthers, and bobcats...would be virtually extinct around settled areas.[26]

By that time, shortages of game had already precipitated drastic changes in the traditional lifeways of indigenous peoples. In fact, livestock had increasingly displaced native game: "as farms replaced forests, cattle, sheep, and goats supplanted deer, elk, and buffalo."[27]

Moreover, with the cash-based economy of European trade goods came the temptation to forfeit the traditional subsistence practices that had always sustained indigenous peoples abiding by ethical principles of reciprocity with the natural world. Growing pressure on resources—especially overharvesting—could quickly disrupt established subsistence practices while also undermining age-old indigenous ethical and spiritual beliefs associated with the hunt. In the second novel, the character War Club, the oldest man to have survived the march, recollects the onslaught of the fur trade in the old territory of North Carolina throughout his childhood. As he is dying in Indian Territory many years later—between visions that mix Cherokee metaphysics with Christian mysticism—he recalls when white traders precipitated profound changes to wildlife populations in the woodlands of the Southeast:

> He remembered as a child hunting for deer with his father. It was what the traders wanted—deerskins. War Club knew the magic song to call the deer to them.... Sometimes he thought he heard the deer sing their death song. He knew the deer had a language. He remembered a hum in the air during the skinning, the dismembering. He remembered the deerskins piled up for the traders. The traders let the Cherokees keep the meat. All they wanted was the skins. He thought sometimes they would clear the woods of deer.[28]

In this manner, furbearers were routinely extirpated as frontier settlement swept westward across, and ultimately beyond, the ancestral homelands of the Cherokee. Moreover, as the fur trade devastated populations of game the tribe had held sacred, a resulting clash of values fundamentally disrupted Cherokee traditions of spiritual leadership. When traders and landholders began to assume status and power in their communities, leaders charged with perpetuating ancient traditions and rituals were subverted. Such epic changes are situated at the cusp of cultural and environmental history.

By the nineteenth century, in fact, many of the Cherokee had assimilated at least partially to European-style gardening and farming. Conversely, European settlers to the region—like the indigenous peoples they would displace—were equally dependent on growing corn, "that grain of American origin so crucial to the subsistence of both Indians and colonists."[29] For tribes throughout the region whose relationship to this revered crop was a far more ancient one, corn remained the very staff of life. Accordingly, the cultivation of corn is a recurring motif in *Pushing the Bear*—a central symbol of Cherokee identification with the land. "We were the land," Maritole declares in the novel's opening pages, "the cornstalks were our grandmothers.... Their voices were the long tassels reaching the air. Our spirits clung to them. Our roots entwined."[30] Ceremonies associated with planting as well as with hunting renewed a sacred relationship with nature predicated on reverence and reciprocity—rituals to ensure a bountiful harvest such as singing to cornstalks and the Green Corn Dance, which

Silver characterizes as "an elaborate ritual of purification and celebration associated with the ripening of maize."[31] As a character in *Pushing the Bear* says, "The fields we are used to were worked for generations. They had been blessed."[32] Those who endured relocation, however, were soon to lose this complex legacy of cultivation. Perhaps above all they would yearn for the small fields that their elders and ancestors had cleared and planted with corn. Indeed, in Indian Territory they would again pray over the newly cleared fields that they planted as they had in the old territory, imploring the land to "become holy with corn."[33]

The loss of ancient ties to the landscape that had shaped Cherokee traditions and beliefs over many centuries were twofold—ecological and cultural. Indigenous belief systems throughout North America were predicated on a reverence toward the natural world, consecrated for many centuries in particular landscapes. Glancy invokes just such a spiritualized Native American view of nature in the first novel by retelling a Cherokee "legend" of "the Medicine Lake of the birds and animals." This narrative expresses a theme of restoration: animals wounded by hunters are immediately healed after plunging into the water. Could such medicine bring back former levels of natural abundance once game has been decimated by the fur trade? "If a hunter comes near the place he would know it by the whirring sound of thousands of wild ducks flying over," Maritole tells a sympathetic soldier.[34]

> But upon reaching the spot the hunter would find only a dry flat without birds or animals unless...he first sharpened his spiritual vision by prayer and fasting.... Because it's not seen, some people think the lake has dried up, but this is not true. To anyone who kept watch and fasted through the night, it would appear at daybreak. In the water all kinds of fish swam upon the surface, and overhead there are flocks of ducks and birds, while around the shore, bear tracks cross in the four directions.[35]

Indeed, the lake's magic raises the prospect of a broader redemption beyond momentary abundance—but only on the condition of performing prescribed rituals such as fasting and prayer. By extension, could overall environmental decline—particularly widespread extirpation of wildlife since the settlement of whites—be mitigated by such an expression of reverence for the healing powers of Earth?

The Cherokee, like many Native American peoples, renewed a sacred relationship to nature through ceremonies that accompanied subsistence practices. Hunters, for example, prayed to the spirit of deer—expressing gratitude and asking forgiveness—to help ensure successful hunts in the future. As Silver explains: "By asking a dead deer's forgiveness, celebrating the maturity of corn, or appeasing the spirit of ginseng, Indians could take what they needed from the forest without violating or compromising that basic spiritual relationship between people, plants, and animals."[36] Similarly, conjurers would consult the spirits of certain plants that according to oral tradition had knowingly taken on medicinal qualities, actually addressing them as sentient beings when gathering them. Such religious practices characteristically entailed an evocation of spirits, celestial and terrestrial, as Glancy depicts in the case of the Cherokee: "those who worshiped the old way...called to the sun and moon. They called to the wolf, the blue jay, the mulberry."[37] Over time, indigenous beliefs increasingly coexisted with church precepts accepted by those Cherokee who converted to Christianity, whether the two worldviews were perceived to be irreconcilable or blended instead. *Pushing the Bear* portrays how the Cherokee traveling westward along the Trail toward Oklahoma attempted to listen in new places for familiar animistic voices, both wildlife and plants: corn, trees, and indeed the forest as a whole.

In the wake of the Trail of Tears, the Cherokee sought to reestablish their spiritual connection to nature. The cherished relationship to ancestral homelands was soon only a memory. No amount of prayer, song, dancing, or conjuring seemed to have any power to avert the tragedy that had befallen them. No sooner had the Cherokee begun staking claims to farmsteads in Indian Territory than the process became greatly complicated by the presence of other groups of Native Americans such as the Osage and Creek, who had been relocated there previously. In addition, Old Settler bands of Cherokee that had moved west twenty years earlier after agreeing to sell their farms in the East had already cleared and planted some of the parcels reassigned to the new arrivals. Some of them wondered whether spiritual connections to the new land already established by the Old Settler bands over the last two decades might extend to those who had only just arrived. Yet in the new territory, parcels for homesteads were deeded as personal property held by individual families, fundamentally transforming ownership of once communal cornfields. Consequently, the only way to reembrace their cultural inheritance, the Cherokee realized, would be to reshape traditional rituals in relation to a new place, in the hope that their indigenous identity would again be made manifest by the land itself.

Relocation—like the initial arrival of white settlers—had precipitated tumultuous change in the lifeways of the Cherokee, making their very survival precarious. In both novels, we glimpse their predicament from the perspective of an actual historical figure, Oo-no-du-tu, the converted Indian minister Reverend Bushyhead, born in Tennessee in 1804. As a child he had witnessed pervasive landscape change in the region as white settlement spread across the Southeast and forests had been logged off:

> A nearby county, Rhea, had nearly doubled in size with the annexation of part of the Cherokee nation. He remembered the rows of hills and mountains they had left. Sometimes he thought they called to him. He thought of the streams and creeks.... He could almost hear the moccasins of a runner through the woods. The sound of the first splots of rain. Birds talking. A deer running on its hind legs. It was a sign that things would change. The settlers annexed their country, the yellow red and yellow brown soil, the fir trees and oaks, words heard in the streams, the gullies and ravines. Bushyhead remembered when civilization was a few wagon ruts and footpaths in the woods.... He remembered logs fallen on the floor of the woods. Felled logs. Axes for cutting trees. Planes for plank flooring.[38]

After the Cherokee arrival in Indian Territory, Reverend Bushyhead characterizes the tumultuous transition in a sermon: "They had once been hunters, who became farmers, who had been removed from their land. Now they were hunters again, poor, illiterate, lost. They were trying to become farmers again. Meanwhile, they hunted for food to eat. They hunted for meaning—for cause to continue struggling."[39] Wildlife initially encountered in the region included a variety of desirable game: fowl such as pheasant, quail, and turkey; deer, and furbearers such as badger, beaver, and bobcat.[40] Since such game populations were increasingly subject to extirpation, however, surviving relocation would ultimately require adopting European lifeways: they would soon resume animal husbandry, raising chickens, cows, and pigs, as many had already begun to do on the eastern frontier.

The landscape of Indian Territory differed dramatically from that of their ancestral homelands in North Carolina. Trees were fewer and comparatively small—and the stands were relatively sparse compared to the denser forests of southeastern woodlands. In addition, planting prairie grasslands entailed the brutal labor of hacking fields from untilled turf, breaking prairie sod and severing deep and resilient roots, and plowing ground choked with stones. Of equal concern was the fertility of the soil. Could it sustain their traditional crops, such as beans, corn, pumpkin, and squash? Before even beginning planting with the few seeds they had managed to smuggle along, those just arriving tried to determine promising places to stake claims by relying on the voices of these seeds as well as the land itself: "Have you felt the place that calls the seeds into to ground," Knobowtee asks Anna Sco-so-tah. "I feel the seeds struggling to get into the ground," she said. "*Here*, I hear the spirits saying."[41] Clearly, the animistic belief in the volition and agency of plants and animals in the wild had carried over into the Cherokee's cultivation of corn. In Indian Territory, they went on listening for the voice of the new land as they planted and prayed for their crops to prosper once more. Traditional planting ceremonies that had not survived the journey would have to be reinvented in relation to the new landscape.

For many who endured the Trail of Tears, cultural renewal began with cultivation itself: "We will be ourselves," Maritole's brother Tanner reasons, "when we eat our own corn again" rather than government rations and store-bought commodities.[42] For corn to flourish in the newly cleared fields, however, they would need to reestablish a culturally mediated relationship between the plants and the land—one that Glancy portrays in terms of language and voice: "The stalks had to learn to speak to the soil.... The corn was speaking to the ground. It was telling the stones to move over. It was telling the ground it had come to the new territory."[43] Historically, the Cherokee ascribed agency even to inanimate elements of the landscape and listened patiently for such voices. With the death of so many elders along the Trail, however, a number of essential planting rituals were lost, along with those the survivors had forgotten. Conjurers charged with performing the incantations at planting, for example, were absent: "Many of the orators did not survive the removal trail. The Cherokees had to plant fields for their survival. They could not wait for an orator to come to the farm to plant the blessing.... There were stories of orators getting lost, trying to find their way to the new farms. There were stories of them disappearing on the journey."[44] Still, surviving elders attempted to reestablish communication with the new land before planting, "speaking to the soil, telling it to open to the corn, telling it stories."[45] "Here is what the rocks say," declares one elder, Anna Sco-so-tah. "The old fields will speak to the new. They will help us replow."[46] Above all, it was the People themselves who had been utterly uprooted and were now to be transplanted, as Maritole recounts: "Our land in North Carolina was gone. We could never go back, any of us. We would never be the same.... Was there anything to replant in the new land? What would come of us?"[47]

After their arrival in Indian Territory, the Cherokee doggedly seek to resume traditional spiritual practices in the unfamiliar land. At first the cacophony of local wildlife seems utterly unintelligible to them: "The old territory had been noisy with birds and animals.... The noise in the new territory was new.... All were talking, but none were listening. All were telling their stories—the frogs at the creek, the ducks, the whippoorwills, the owl, all of them, woven together."[48] They perform time-honored ceremonies to restore their connection to the natural world. The seven clans gather to assemble ceremonial arbors:

> They had to have the stomp dance. The life of the people was in the ceremony. It was an act of resilience—an act of alignment with old customs through which the old power came.... They heard the old rhythm of the land. They performed their ceremonies until they felt the magic that made things work. They had to unpack the old ceremonies after the removal trail. They had to find out how to re-ceremony the old ceremonies in a new place.... The ancestors and the old ways gathered in their dancing. The old language became one with itself.[49]

Only then might the voices of the natural world become recognizable as "the birds and animals sounded as they should" once again.[50] In their ancestral homelands to the east, the Cherokee had readily interpreted and heeded such voices. By contrast, surviving in the new environment of Indian Territory entailed reorienting themselves, practically in terms of subsistence, and spiritually in terms of cultural identity.

Traditional rituals the Cherokee had historically performed in their ancestral homelands were ultimately reconceived. After arriving in Indian Territory, as Glancy describes, the conjurers "realigned the chants and ceremonies that had accompanied them on the removal trail.... It was part of their magic to move. To split. To retrieve. To rethink."[51] They would need to learn to speak to the new land and to communicate with stones. "The conjurers told the rocks to move aside as they plowed.... They spoke their way into the new territory—rendering their words in accordance with the old ways.... They had carried the knowledge that had made their way since the beginning. They worked with their remade language, calling it from its old form, rethinking it, recreating it...How things had been...would continue to be."[52] The new land would need to be transformed, they understood, and in the process, the People would be transformed as well.

As Maritole and her fractured family cross Arkansas toward the border with Indian Territory, approaching the end of the Trail, she suddenly hungers for the old stories, and begins recounting one she remembers, a creation story, as her father dies beside her:

> "At one time all things were in the sky.... All living things spoke the same language, so they understood each other. But we misused this privilege and were stricken deaf to the talk of animals and birds.... As the birds and animals grew in numbers, the sky became crowded until there was danger of being pushed off.... Finally the water beetle was sent to see if he could find a place where the animals could live.... He dived down under a large body of water and found mud.... He brought the piece of mud to the surface.... The mud began to grow until after a long time it covered much of the surface of the water.[53]

She realizes that the Cherokee, dispossessed of their ancestral homelands, face a similar fate: "We're like the animals in the sky pushed off to find a new place."[54] Listening for the voices of ancestors, she wonders just what fragments of Cherokee culture have survived the journey, and especially the fate of the tribe's sacred fire. Finally, she is reminded of the legend of Medicine Lake that heals wounded animals—that place of redemption—revealed only to those who practice traditional spiritual rituals:

> Always now in the dark I heard the voices of our ancestors. They said to keep walking. The spirits who had held the raft across the huge Mississippi River told me the road would come to an end and I would see the new land.... Sometimes stories didn't seem to make any difference, but now I wanted to ask my father for a story again.... Suddenly I felt the magic of the stories again.... Inside the Keetoowah fire burned and some of our legends lived. The thought of the magic lake ᎠᏔᎦᎯ would always heal.[55]

To the present day, Cherokee identity is bound up in the legacy of relocation and a sense of environmental history. Indeed, Cherokee cultural heritage entails having endured the travails and trauma of removal. In *Building One Fire*, a recent treatment of traditional and contemporary Cherokee art, co-authors Chad Smith, principal chief of the Cherokee Nation, and Cherokee-Osage scholar Rennard Strickland close by expressing indigenous identity by both perpetuating and reshaping tribal tradition. They acknowledge the valiant efforts of past generations that helped ensure that Cherokee identity would endure:

> Our ancestors are always close.... They are the embers that flow at the core of the sacred fire. When those embers were brought on the Trail of Tears from our ancient homeland in the East they ignited the past in the new lands of the West.... Our forebears, whose lands were stripped from them, whose children and grandparents starved along the road of removal, and who amidst unimaginable despair built their new nation in the virgin forest of Indian Territory.[56]

Indeed, relocation to the "virgin forests" and grasslands of Indian Territory amounted to dramatic landscape change. Time after time since the arrival of whites, indigenous peoples throughout North America have faced epic environmental disruption of various kinds that compromise vital natural resources, such as habitat for wildlife, by incessantly encroaching on the hunting grounds they had relied on for sustenance over millennia. Glancy's novels offer an intimate glimpse of what was lost when they were forced to abandon their ancestral homelands: a profound legacy of historical, cultural, and spiritual significance. Moreover, the novels demonstrate how Native American cultures have been powerfully shaped by a sacred relationship to place and reverent communication with nature. For the Cherokee, as for many of the continent's other indigenous peoples, environmental history and cultural identity have been—and always will be—inseparable.

NOTES

1. Glancy, *Pushing the Bear: After*, 187.
2. Ibid., 193.
3. Ibid., 193, 188.
4. Glancy, *Pushing the Bear: A Novel*, 75.
5. Ibid., 47; spoken by Knobowtee.
6. Glancy, *Pushing the Bear: After*, 188.
7. Ibid., 188.
8. Ibid., 108.
9. Ibid., 61.

10. Ibid., 6.
11. Ibid., 7.
12. Silver, *New Face on the Countryside*, 41–42.
13. Glancy, *Pushing the Bear: A Novel*, 128.
14. Glancy, *Pushing the Bear: After*, 3.
15. Glancy, *Pushing the Bear: A Novel*, 220.
16. Ibid., 158.
17. Ibid.
18. Ibid.
19. Glancy, *Pushing the Bear: After*, 17.
20. Glancy, *Pushing the Bear: A Novel*, 232.
21. Silver, *Mount Mitchell and the Black Mountains*, 32.
22. Silver, *New Face on the Countryside*, 5.
23. Silver, *Mount Mitchell and the Black Mountains*, 46.
24. Glancy, *Pushing the Bear: A Novel*, 189, 206.
25. Ibid., 75.
26. Silver, *New Face on the Countryside*, 186.
27. Ibid., 187.
28. Glancy, *Pushing the Bear: After*, 104.
29. Silver, *New Face on the Countryside*, 5–6.
30. Glancy, *Pushing the Bear: A Novel*, 4.
31. Silver, *New Face on the Countryside*, 41.
32. Glancy, *Pushing the Bear: After*, 59.
33. Ibid., 37.
34. Glancy, *Pushing the Bear: A Novel*, 125.
35. Ibid.
36. Silver, *New Face on the Countryside*, 42.
37. Glancy, *Pushing the Bear: A Novel*, 86.
38. Glancy, *Pushing the Bear: After*, 156–57.
39. Ibid., 81.
40. Ibid., 77.
41. Ibid., 22.
42. Ibid., 78.
43. Ibid., 136.
44. Ibid., 48.
45. Ibid., 143.
46. Ibid., 46.
47. Glancy, *Pushing the Bear: A Novel*, 113.
48. Glancy, *Pushing the Bear: After*, 14–15.
49. Ibid., 96–97.
50. Ibid., 96.
51. Ibid., 160.
52. Ibid., 161–62.
53. Glancy, *Pushing the Bear: A Novel*, 211–12.
54. Ibid., 212.
55. Ibid., 210.
56. Smith and Stickland, *Building One Fire*, 221.

9

Time's Horizon

Elizabeth Dodd (b. 1962)

This is the world that dies
and comes back, that dies and goes

on or this time doesn't.

—Elizabeth Dodd, "Dieback"

Elizabeth Dodd has emerged over the last two decades as a major voice among contemporary midwestern literary naturalists, publishing three volumes of essays as well as two collections of poems that form a cohesive body of work probing our relationship to place. Many of her essays describe journeys that represent personal pilgrimages to view archaeological sites in the Southwest, Rocky Mountains, and Pacific Northwest as well as the storied Paleolithic cave paintings in France. She finds the tug of the ancient on the imagination to be irresistible, reading each landscape in terms of both environmental and cultural history. In the process, she reflects on the history of language itself: the etymology of words in relation to geographical context, as well as the origins of individual languages, especially how those of the indigenous peoples of North America hold "ancient histories of migration and assimilation."[1] In addition, each of her books offers keen perceptions of the natural world, while frequently contemplating the vast expanses of time associated with geology, evolution, and anthropology. In its repeated return to these themes, her writing might be likened to Loren Eiseley's. Indeed, they share a fascination with enigmatic and often animistic aspects of the distant past. Critic Susan N. Maher credits Eiseley with "inventing a new form of nonfiction essay, one that could embrace the depths of time and geology…within a literary line of descent that commences with Thoreau... illuminating cross sections of natural, national, and personal history; [exhibiting] a reflective tenor of loss that leads to the retrieval of lost worlds and honors vestiges of the past."[2] His work has been linked to the pantheon of American nature writers "who imaginatively delve into environmental explorations, forging connections to ecological and evolutionary dynamics... close observation of phenomena, keen awareness of environmental processes, appreciation of species and their habitat, aesthetic engagement with the natural world, philosophical reflections, and a recognition of the deep kinship between humans and the biosphere."[3] I would contend that Dodd's work follows in this lineage while extending the hybrid narrative genre, deploying the same range of rhetorical means to address a consistent constellation of themes but with an added emphasis on the history and perspectives of the continent's indigenous peoples.

Having grown up in the Appalachian foothills of southeastern Ohio, Dodd's interest in studying the natural world was first inspired by experiences exploring the second- and third-growth woodlands near her childhood home. She recollects "numinous moments that bring the outer to the inner realm, unstructured by will," such as an early memory of listening with her family on a forested ridge to the hauntingly beautiful call of a hermit thrush.[4] Such recollections suggest a spiritual aspect to our relationship to nature, a quality that much of her writing conveys. Like several of the literary naturalists whose sensibilities originated in the region—including Paul Errington, Paul Gruchow, and Scott Russell Sanders—she has a deep regard for environmental history and landscape change. Her underlying aim is "to look at time itself," as she declares in the opening of her third collection of essays, *Horizon's Lens*, to examine "a biological legacy from our evolution... to feel the pulse of our existence synchronize with deeper cycles."[5]

Even as an adolescent, Dodd had begun to read local landscapes in terms of environmental history, which she perceived as "a kind of historic frontier where we examined the hillsides for bits of history: hidden pathways of old roads, former homesteads, the current communities of maturing hardwood."[6] Wildlife had been trapped out long before, and many of the original homesteads had been deserted. The land had been cut over and cleared, but farms were abandoned here and there, leaving a patchwork forest of varying ages. She and her younger brother would stumble upon the foundations of abandoned farmsteads in the woods: gardens that had gone feral, patches of daffodils that had naturalized. She characterizes the woodlands that remain after several waves of settlement and development as "second- and third-growth forests that had reclaimed southeastern Ohio from the nineteenth-century attempts at subsistence agriculture on steep clay slopes."[7] During her college years, she hiked wooded hillsides of Ohio that showed unmistakable evidence of clear-cutting, with native hardwood forests replaced by tree farms planted with faster-growing pines.

In the Appalachian landscapes Dodd explored during her youth, natural resources had been relentlessly exploited by outside interests for generations. The legacy of coal mining was devastating: toxic tailings and hardpan dirt where nothing grew. Like Midwestern poet James Wright, she recoiled from the destruction wrought by extractive industries such as strip mining, which fouled waterways and often left entire landscapes impaired ecologically and aesthetically. Eiseley described late in his life the seemingly benign yet ubiquitous blight of "the restless advance of suburban housing" that led to the extirpation of species in his Pennsylvania neighborhood, whether the chipmunks displaced by rats or "the last land tortoise [that] had fallen victim to the new expressway. None of his kind any longer came to replace him." An ancient order was obliterated incrementally, one subdivision after another.[8]

Tom Lynch recounts how Eiseley had a keen, lifelong appreciation for environmental history, reflected in the poem "No Place for Boy or Badger," which "laments the destruction of the woods and fields on the outskirts of Lincoln that were his boyhood haunts and their replacement by suburban development.... The rich diversity of nature has been replaced by monocultural conformity."[9] He recognized the profound formative impact of intimate acquaintance with the natural world that was at that time being championed by the Nature Study Movement, sweeping across schools in this country in the early decades of the twentieth century. Indeed, his lifelong grasp of environmental history is encapsulated in the closing paragraphs of *The Night Country*, symbolized by a single cottonwood tree he had planted as a child beside his father. That sapling had grown in his imagination for some sixty years, lodged in memory, until he returned to the site to find it absent: a poignant personal reminder of the erasure of natural landscapes. "This was more than an animal's attachment to a place," Eiseley writes. "There was something else, the attachment of the spirit to a grouping of events in time; it was part of our mortality."[10] This personal account of a heightened awareness of place—and its relation to personal identity—might be likened, albeit in microcosm, to indigenous peoples' cultural memory of traditional homelands, reflected in the mythic and historical narratives of oral tradition as well as the spiritualized subsistence practices of hunting and harvest.

Dodd likewise appreciates how profoundly childhood landscapes had shaped her. She describes a recent return visit as "a kind of pilgrimage to a tiny stand of tallgrass prairies in the upland forest of southeast Ohio.... In such a homecoming, I'd be again on the land, in my skin, coming into the country of the self."[11] The historical significance of the spot stems from its status as a remnant of what had been a more pervasive ecosystem in the region prior to the last ice age. She recognizes, moreover, that this island of prairie is made especially precious by being isolated, rare, and fragile, reflecting her passion for perceiving place and reading landscapes in terms of deep time. Such essays also exemplify her fundamentally interdisciplinary stance as a naturalist, as she effortlessly shifts focus between ecology, geology, and the history of species extinction:

> This little holdout of tallgrass has, indeed, held on—researchers say the soil records a grassland community that existed here at least some thirteen thousand years ago. It's a tiny portion of the Pleistocene world once grazed and ranged by vanished animals that lived south of the ice sheet: the giant ground sloth, stag moose, musk ox, mammoth and mastodon.[12]

Attuned to historical aspects of ecology when afield, Dodd is equally attentive to traces of landscape change in the past found in the written record. She seems just as comfortable discussing *Beowulf* or a thirteenth-century Icelandic poem as the observations of a settler in the Midwest at the dawn of the nineteenth century. In 1804, for example, a Quaker named Joseph Gibbons toured Ohio looking to relocate his family to a place where natural resources might enable homesteaders to prosper. In his journals he exclaimed at the gargantuan size of virgin timber—trunks reaching forty feet in circumference—as well as the biodiversity of the forest. He cataloged five types of oak alone; other nut-bearing species such as beech, chestnut, hickory, and walnut; and fruit-bearing varieties, such as cherries, crab apple, mulberry, and plum.

What interests Dodd in this portrait of midwestern forests preceding white settlement is the glimpse that Gibbons's journal provides of ecosystems that once dominated the region. Ultimately, she seeks to square such textual records with her own impression of primordial forest remnants still standing in our time. She recounts the environmental history entailed in

> the great woods, the eastern deciduous world that preceded European contact. Researchers believe that 95 percent of what is now Ohio was once covered with such forest.... Despite Indian rearrangements of the landscape—farm fields, villages, burned openings to provide grazing land for game—a forest ecosystem once dominated the land east of the Mississippi, now only glimpsed in miniature enclaves, little islets of individual trees rather than wide watersheds of woodland community.[13]

Paddling through swamps along the Black River in North Carolina, Dodd is enthralled by the antiquity of stands of bald cypress there, passing stumps up to three feet in diameter recently felled by a devastating storm. Imagining how such a scene could come to be, she stands in the hollow of a monumental stump, pondering the immensity of time, what Eiseley has described as "cherishing again [the] dream of the Adamic Eden...pursued and lost in the green forests of America."[14] Reading William Bartram's account of exploring the region in 1773, fully a hundred years before such boggy forests were first logged, she discovers that he too marveled at the massive cypresses—up to twelve feet in diameter. Contemporary ecologists have determined from stumps in this remaining forest remnant that the age of some trees there had approached two millennia. How are we to comprehend such antiquity? Dodd contemplates just such questions in the lyrical close of the culminating essay: "So the imagination has leaped again," she writes, "across time and distance, taking me outside myself, beyond myself. Isn't this what we want when we reach toward the spiritual realms that seem to lie both within and beyond us?"[15]

Dodd has also witnessed the antiquity of human culture in the region, such as when visiting the thousand-year-old Great Serpent Mound in southern Ohio, the longest monument of its kind on earth. She realizes that this is precisely the type of site on the cusp of cultural and natural history that so often captures her imagination. In fact, a lifelong fascination with such ancient traces has led her to seek out and contemplate rock art throughout the American Southwest as well as the renowned Paleolithic cave paintings in France. Archaeologists speculate that the mound's orientation carries astronomical significance, perhaps corresponding to the sun's position at solstice. If that was its function, she wonders, what then are we to make of its *meaning*? She conceives of its "ritual role in conceptualizing the rhythm of the seasons. The mound becomes a physical place to keep watch for time's passage and to keep faith with its underlying pattern."[16] Whether at Chaco Canyon in New Mexico, the Great Medicine Wheel in the Bighorn Mountains of Wyoming, or even the main temple at the Mayan ruins of Chichen Itza in the Yucatán, such celestial "predictor sites" typically align the annual peregrination of the sun with agricultural calendars, signaling times to plant and perform related rituals and ceremonies.

Dodd's accounts of disparate landscapes elsewhere are still informed by a keen appreciation for environmental history derived from a lifetime in the Midwest. In the American Southwest, for example, she identifies personally with Georgia O'Keeffe's paintings in terms of the topography they portray. O'Keeffe had begun her career painting Virginia's forested hills and ridges before first relocating to Texas and eventually to New Mexico. Having moved west herself, to Kansas, Dodd had also experienced a fundamental reorientation to the region's distinctive landforms. Adjusting to the grasslands had meant learning to see landscape in new ways and to appreciate what she terms "aesthetics of the horizontal.... A relative newcomer to the plains myself, the rolling mesa hills of the tallgrass prairie in eastern Kansas, I was energetically trying to get the sense of the place, to learn the names of plants, the layers of limestone with their sharp nodes of flint, the exhilarating vista of the wide, bright sky."[17] In Kansas, remaining fragments of the once expansive tallgrass prairie also present her with unfamiliar aspects of environmental history, such as the role of fire "as a shaper of the environment, as a part of the deep background of this *place*... [and] a necessary element of forest succession."[18]

Dodd recognizes the biting irony of sharing—albeit inadvertently—in the spoils of Manifest Destiny: "I have a home on a portion of prairie where 150 years ago no white people lived. We are there now, the beneficiaries of that mostly forgotten conquest, and I earn my living teaching the great great grandchildren of pioneers."[19] Wherever she turns in the region are the tell-tale traces of dispossession. Indeed, she realizes, "In the Black Hills, everywhere, the map of language laid across the landscape, the mythology of western expansion, revealed the ironic gone unrecognized."[20] This is the same environmental history that Eiseley distilled in a capsule narrative of landscape change: "the abandoned cabin of stone that lay far up on one hillside. It had been built in the land rush of the last century and then lost to the cattlemen again as soils failed to take the plow."[21] Here is another example of Dodd and Eiseley's shared penchant for contemplating what Maher terms "the internal and cultural landscape of the Plains."[22]

Consequently, Dodd has been led to ponder the cultural legacy of place-names tied to the indigenous peoples who inhabited the tallgrass prairie for many centuries before being displaced by the whites, including the name of the state itself: "There are few signs of Native American settlements...that have been found on Konza prairie. Still it is named for the Konza Indians—or Kaw, Kansa, or Cansez...Siouan-speaking peoples whose tangible, cultural identities have also largely disappeared, leaving their names like clichéd abstraction, place-names steadily losing their elegiac connection with an earlier existence."[23] Like Louise Erdrich's meditations on the Ojibwemowin language and its relationship to place in ancestral homelands in *Books and Islands in Ojibwe Country*, Dodd posits a direct connection between the two: "language inscrib[ing] an inheritance of inhabitation, an acquaintance with place that merged your own childhood discoveries with intimacies acquired over generations."[24] They share the conviction that indigenous language and landscape are inextricably linked. Indeed, a single region such as the hill country of Kansas could conceivably over time have given rise to multiple tongues. She imagines "word-maps laid out on the prairie landscape over centuries" as necessarily combining "Siouan, the speech of the Kansa; [and] Caddoan, the speech of the Pawnees."[25] Yet such mappings of dialect to landscape are not static but dynamic, as in the case of tribal migration away from ancestral homelands. The oral tradition and "mythic imagination" of the Pawnee, for example, draw on cultural memory of the southern plains before their arrival on the tallgrass prairie of Kansas and Nebraska.

Similarly, Dodd detects the echoes of both cultural memory and environmental history when traveling through the Black Hills on the Flume Trail, dedicated for the hundredth anniversary of South Dakota statehood in 1989: "Along the way place-names call elegiacally across the distance of not much more than a century: at the northernmost tip, Bear Butte Lake trailhead; at the southernmost, Elk Mountain. The large mammals are mostly gone, although a furtive herd of about 200 elk live in Wind Cave National Park to the south, along with a slightly larger herd of bison."[26] For Dodd, the absence of Native Americans today in their traditional homelands is palpable. Standing beside a trace of the Oregon Trail—with its legacy of white migration and tribal dislocation—she recites a litany of indigenous peoples who would have certainly once traveled there: "where countless Indians—Osage, Kansa, Potawatomie, Chickasaw, and others—would have camped decades, centuries before white people arrived to carve their names and dates on the slabs of rock at my feet."[27] She alludes to a trace of Indian ancestry in her predominantly European heritage: "Pale skin obscures... our family's Native American legacy—a bloodline that has lost contact with the cultures it once propelled. Mine is an uncertain inheritance." It is an ancestry apparently too subtle for rudimentary genealogical DNA testing to detect.[28] "My own grandparents are gone," she explains, "and with them some of the hints and whispers of my bloodline's native connection with the continent. Choctaw—late arrivals to the southern plains, some of my father's people; they spoke Muskogean, a language of the Southeast. And my mother's people—well, their story's been garbled through generations. Algonquin? Athabaskan?"[29]

Ever attentive to cultural memory, Dodd learned the word for buffalo in a variety of indigenous languages. She recounts the epic decline of bison, beginning with their historic range and culminating with their fate in the state of Kansas:

> Once they grazed in the grassy stretches of Kentucky, kept open, most likely, by Indians who set fires there to provide those tremendous game beasts with pastureland. Bison lived throughout the East: Pennsylvania, West Virginia, Tennessee, western Georgia, the Carolinas, and, farther west, the plains of Indiana and Illinois. In the 1700s they were found—and killed—along rivers I knew growing up: the Kanawha, the Scioto, the Muskingam, and of course the Beautiful River, the Ohio.... The last bison was killed in Ohio in 1802. By 1834 the animals were largely absent from eastern Kansas; in the next 40 or 50 years they would disappear from the rest of the state.[30]

The concerted and ultimately successful campaign to rescue the species from the brink of extinction represents, a century later, a foundational narrative of contemporary wildlife restoration.

Elk provide a ready corollary, as Dodd describes: "Before European settlement, North America may have been home to as many as 10 million elk; by the end of the nineteenth century they had been extirpated nearly everywhere... and they numbered only about 50,000.... By the end of the [twentieth] century their numbers had reached 783,000."[31] While current herds are largely associated with mountainous regions of the Far West, she reminds us that the historic range of the species was once far greater, stretching from the woodlands in the East to the plains and prairies of

Kansas. She maps their distribution as once extending to "the northern half of states like Alabama, Georgia, and Mississippi. The arid Southwest...the Pacific Northwest...[and] the very westernmost rim of Massachusetts."[32] She notes that the ecology of each region influenced the physiology of the animal, producing several regionally distinct subspecies, such as the Merriam elk of the arid Southwest. For the ongoing processes of natural selection are ubiquitous, as Eiseley recognized: "Partial adaptations, transitions from one environmental world to another, animals caught in slow motion half through some natural barrier are all about us."[33] Indeed, he believed that evolution might even be conceived as a cumulative "history of adaptation to place."[34] Dodd speculates, moreover, that specific habitats as well as landscape change may also alter animal *behavior* in fundamental ways, ranging from feeding to migration.

Dodd sounds a note of hope—"In contrast to the pervasive narrative of habitat destruction, population crashes, extinctions"—at the recent return of species such as the bald eagle that had been locally extinct throughout much of the Midwest.[35] Eagles—long the victims of overhunting, habitat loss, and toxic chemicals—had like many avian species suffered declining numbers and sheer disappearance, resulting in severe reduction of their historic range. While federal protection of endangered species and prudent environmental measures such as banning toxins like DDT undoubtedly contributed to the successful reintroduction in the Midwest, she recognizes that such recoveries are in fact a testament to the resilience of the species. Evidence of adaptability in feeding and nesting—what could even be considered the *creativity* of these birds—has led to reestablishing themselves in suitable riparian habitat along river corridors and adjoining reservoirs.

Dodd is also drawn to contemplate deep time in terms of cultural memory. Exploring rock art sites throughout the American West, for example, she considers their antiquity, ranging from nineteenth-century images made by the Paiute in California's Coso Mountains to those attributed to pre-Clovis Paleo-Indians that scientists date to the Pleistocene over sixteen thousand years ago. Recounting a visit to the Great Medicine Wheel atop a mountain in Wyoming, she describes how the mystery at such archaeological sites is compelling: "the apparent timelessness...the uncertainty of its origin, the antiquity of its presence in the landscape."[36] Could it have been a ceremonial way station once, on the route of an ancient seasonal migration, she wonders. She distills the enigma of the rock art into a single, deceptively simple question: "what is left to us from them."[37]

Their purpose, she decides tentatively, might be "to inscribe perception into geological fact. Personal impermanence is a given."[38] Yet remarkably, Dodd reports, shamanic rituals linked to specific sites have been perpetuated until recent times, at least in isolated cases, such as by the Paiute and Shoshone, who have continued to retrace the figures of preexisting glyphs.

While examining the rock art in the ruins at Bandelier outside of Santa Fe, New Mexico, she describes what it is she longs to find: "Sometimes I felt a wistfulness for the long-dead artist who painted some pictograph on a sheltered rock face or who pecked a bold petroglyph into the high, visible cliff face. What could such a person teach?"[39] Such questions tantalize the imagination. While some images are abstract forms and geometric shapes such as spirals or diamonds, others appear to be stylized representations of natural symbols such as animals favored as game, or lightning and rain, or even the unmistakable tracings of human hands. Yet, even representational figures readily recognizable today such as animals might have once also carried symbolic, ceremonial, totemic, or even shamanistic connotations. Still other petroglyphs suggest characters in mythopoetic narratives from oral tradition, such as creation stories etched on stones at the base of Spirit Mountain on the edge of the Mojave Desert. Not every image can be identified with certainty, especially when the cultural traditions that gave rise to them have been thoroughly disrupted. Those representing supernatural beings derived from dream or trance, for instance, sometimes prove indecipherable to the modern eye. Indeed, once the stories are no longer told and rites no longer performed, we are left to wonder how much has been lost. Like other literary naturalists such as Paul Gruchow and Scott Russell Sanders, Dodd recognizes the epistemological dilemma of interpreting such ancient images in the absence of the cultural knowledge they originally embodied.

At a few sites, petroglyphs evidently represent the route of ancient journeys, perhaps of tribal migrations. Such geographic mappings are found throughout North America, including in the Midwest. An elaborate map of trading routes of the Mississippian Mound Builders, for instance, has been deciphered on a riverside stone associated with the Cahokia complex in western Illinois.[40] For Dodd, such texts inscribed on the landscape, though sometimes indecipherable, beckon to the historical imagination: "The urge to efface the modern self and imagine—recover—the humanity of earlier peoples," she supposes, "must be a part of what draws us to museums and monuments like Cahokia."[41] She designates the site as "the tallest prehistoric earthen construction in the New World. At the time of the Crusades, in the historic faltering of Europe's Middle Ages, the people of Cahokia prospered, with a city larger than its contemporary London." It functioned as "a kind of cultural fulcrum for the agrarian urbanites who worked and died throughout the region a thousand years ago."[42]

Similarly, visiting the ruins of Anasazi habitations and granaries in the American Southwest, Dodd observes petroglyphs attributed to the Fremont culture some seven to twelve hundred years ago. She recounts how climate change completely disrupted such desert cultures, a cautionary tale in our own era of global warming: "Fremont art arises from centuries of desert subsistence living, generations of dry climate and deeper drought.... Devastating droughts that wracked the Southwest in the twelfth and thirteenth centuries... sweep[ing] the farming peoples from their earlier lifeways, emptying the great communal cities farther east and south, like those of Chaco Canyon."[43] The author is no stranger to Chaco Canyon, having sensed the power of the ruins while camping decades before engaging in the research for the more recent essays set there. An early poem describes ruins the Anasazi left behind in the wake of unrelenting drought, depicting them with archetypal imagery: "The village wall has fallen; / grass covers stones tumbled / from the ancient cistern."[44] "Chaco Canyon, New Mexico" describes the petroglyphs she first encountered there:

> Angular figures, familiar
> from postcard photographs:
> a man, a woman bearing children,
> a mountain sheep, a snake, a spiral.
> ....
> Painted on the rock directly overhead:
>   a hand, a crescent moon,
> a giant star.[45]

Perhaps most striking is the visceral impression it leaves coupled with the mystical response it evokes:

> This is the center of the world,
> the canyon's expanse enclosed
> by cliffs, the horizon encircling all
> there is, and we
> are here, alive.[46]

Such passages reflect a sensibility central to ecopoetics generally, which David Gilcrest describes as "a geological and evolutionary sense of time."[47] Among images of rain showers and snakes, birds and spirit helpers, Dodd senses the same raw, shamanic presence she recognized when visiting still more ancient paintings in the caverns of France, what Eiseley described as "striking three-dimensional art—the art of the brother animal seen in beauty."[48] Singular images, such as the thunderbird, for instance, were pervasive across North America, providing evidence, she contends—much as DNA might—of common ancestry and even clues to human migration in deep time: "The thunderbird is a true cross-cultural element of Native American mythology.... In this metaphor, I take symbol, or mythic figure, as a marker of the cultural journey into the world of figuration, of representation and understanding."[49] Viewing the myriad forms chiseled into rock at Chaco Canyon, she muses on their underlying meaning: "Migration, I think; if the spiral connects with temporality, the birds bespeak seasons."[50] Above all, such petroglyphs suggest to her the emergence of a distinctly human form of expression.

Having visited a plethora of petroglyph sites throughout the American West and viewed hundreds of animal images ranging from birds and snakes to coyotes and bighorn sheep, Dodd is drawn to the idiosyncratic depiction of a mastodon at a remote location on the south bank of the Colorado River in eastern Utah. She is left transfixed by the rock art of ancestral Puebloan peoples of the Great Basin representing species long since extinct.[51] The fact that Paleo-Indians hunted megafauna in North America at the close of the Pleistocene was established irrefutably nearly a century ago. Archaeologists, including Eiseley, have speculated that extinction of major species during the late Pleistocene, and the consequent reduction of available game, correlates with the onset of agriculture—including corn, squash, and gourds—perhaps initially in regions such as the arid Southwest devoid of remaining herds of bison. While calculating the precise age of sites is unfeasible at times, the mastodon image is undoubtedly one of the most ancient records of environmental history in the Americas. "And if it offers the relict, stylized glimpse of an animal icon of the ice age," she concludes, "time's arrow will cross distances unmeasurable by any of the tools we have at hand."[52]

The fabled Paleolithic cave paintings of France add a still earlier evolutionary backdrop, and yet another dimension of environmental history. They depict species that have been extinct since the Pleistocene, such as the woolly mammoth and rhinoceros, as well as the giant Irish elk, calling to mind for Dodd "a list of vanished animals: bears, sloths, mammoths and mastodons, great wolves and cats."[53] As Eiseley observed, such extinctions have been an ongoing force shaping the course of evolution, often in response to landscape and climate change: "The vast majority of the earth's creatures—perhaps over ninety percent—have vanished.... The specialized perish with the environment that created them."[54] In the sweep of evolutionary history, he reminds us, every species, including humans, is mutable and ultimately transitory. Indeed, megafauna of the age included enormous creatures ill equipped to cope with a warming world when the ice sheets retreated. Dodd finds such ancient paintings compelling as an aesthetic and perhaps spiritual expression. These images represent human culture and its relationship to the natural world before the domestication of animals or cultivation of crops. In aggregate, they provide a glimpse, albeit often perplexing, of Paleolithic peoples as well as the great beasts that once sustained them, inspiring their imagination and challenging our own. For, as Eiseley appreciated, "worlds drawn out of nature are human worlds."[55]

Recent archaeological research has investigated the accuracy of ancient cave paintings of animals, both species now extinct and those still extant. DNA analysis of bones found near the caves, those of animals hunted by the same Paleolithic peoples who made the paintings approximately twenty-five thousand years ago, provides evidence of three distinct subspecies of horses, for example, recognizable in the paintings by their varied pelage: brownish bay, black, and dappled. Establishing the anatomical accuracy of such cave art, and thereby its legitimacy in the study of natural history, lends credence to similar images at other sites, including representations of species that have long since gone extinct.[56] Dodd senses the unfathomable temporal dimensions of such images. She has found in these cave paintings what amounts to a profoundly early, Paleolithic record of environmental history, one that suggests the ritualized, metaphysical dimensions of such hunting cultures—arguably a fundamental expression of human religious impulse. The "hunting art" of the Cro-Magnon, Eiseley asserts, presents evidence that "behind all religions lurks the concept of nature."[57] In this view, rituals pervaded by a sense of union with an animistic universe gave rise to art and religion, which he characterizes as nothing less than "the most enormous extension of vision of which life is capable: the projection of itself into other lives."[58] Moreover, he suggests, "Ever since man first painted animals in the dark of caves he has been responding to the holy, to the numinous, to the mystery of being and becoming."[59] Similarly, Barry Lopez observes, "All we do know, staring up at the cave walls at Lascaux, is that the belief that there is more to hunting than killing, and that dying was as sacred as living, was not something that one day just fell out of the sky."[60]

Such cave paintings also reflect a history of gradual landscape change: an inadvertent record of extinction in the wake of glaciation that also maps the development of human expression. Dodd notes that though many of the most celebrated cave paintings date from twelve thousand to eighteen thousand years ago, known as the Magdalenian Period, others of comparable sophistication were completed ten to twenty millennia earlier—evidence of enduring cultural continuity. After viewing the images of animals at Labastide—an array of ibex, waterfowl, bison, and wild Pleistocene horses—she is gripped by the visceral power of the cavern at Chauvet. Beyond the "visual vitality" of its horses and bison, lions and rhinos, are the traces of human use: "small piles of burned wood, tiny artisan's fires… scent of char from 30,000-year-old fires."[61] She had already imagined such a scene in her poem "Tsankawi," published decades earlier: "breathing in the cooler air, the ancient smell / of ritual soot, you could almost lose / your sense of loss."[62]

Dodd leaves this subterranean shrine with a renewed sense of awe and humility: "despite the amazing age of the galleries and the animate images of a lost age, the hints of symbolic meaning that continue to elude us, despite all these hints and recognitions, we are departing."[63] Yet, as Eiseley would have it, such "art devoted to the unseen, to the powers that control the movement of game," were in effect the initial steps toward a host of human technologies that endeavor to manipulate or even dominate nature to the present day:[64] "The mind that once visualized animals on a cave wall is now engaged in a vast ramification of itself through time and space.… Knowledge has given [us] the memory of earth's history beyond the time of [its] emergence."[65] Such sites offer a testament to the environmental history of deep time, as well as to a human impulse that seems at once utterly distant and intimately familiar. Dodd attempts to articulate several dimensions that such images suggest: "Transformational, surely, I think. The figures seem to me to posit the wonder of being, the metaphorical or spiritual complexity of being-in-a-world-of-being. The wonder of the body in the flex and extension of perception, even perception of the self as other, suddenly here.… The image, whatever its import, is fused with location."[66]

For Dodd, reading landscapes of the American Midwest for evidence of environmental history ultimately entails coming to terms with the region's geological legacy: "On the tallgrass prairie, in winter, I try to imagine the steppes of another era," she writes, "the Pleistocene world when, in the last glacial maximum, grasslands wrapped Europe's preglacial landscape, from the Atlantic coast of modern France to the deep interior of Eurasia."[67] Here we again detect echoes of Eiseley, who described such geological history in remarkably personal terms, as if it is directly implicated in the circumstances of his own birth: "I was...conceived in and part of the rolling yellow cloud.... That cloud has been blowing in my part of the Middle West since the Ice Age...impenetrable as it was when our ancestors first emerged from it on the ice-age steppes of Europe, or when they followed the bison into its heart on the wide American plains."[68] He was convinced, moreover, that the vestiges of ancestral memory lingering since the ice age underlie our own visceral reaction to Arctic environs and the diminishing light of autumn in the far north (or even Thoreau's awe in the face of what he perceived as primordial nature on a mountaintop in Maine), and he depicted the temporal and spatial dimensions of such scenes with striking immediacy, such as in his essay "Man the Firemaker."

> Some 15,000 or 20,000 years ago. The last great ice sheet lies across northern Europe and North America. Roving on the open tundra and grasslands below those ice sheets is the best-fed and most varied assemblage of grass-eating animals the world has ever seen. Giant long-horned bison, the huge wild cattle of the Pleistocene, graze on both continents. Mammoth and mastodon wander.... The grasslands are alive with seemingly inexhaustible herds of game.[69]

For both Eiseley and Dodd, this geographical reach is compounded by the grand expanse of time[70]—an epoch when geological and climatic upheaval triggered a cascade of environmental consequences that ultimately resulted in the disappearance of many species, while dictating the distribution of others across the continents. In fact, many great migrations that survive to the present day are predicated in part on the advance and retreat of glacial ice. Such musings about historical extinctions triggered by climate change seem hauntingly prophetic in light of the escalating ramifications of global warming today. The extent to which human activities have influenced these processes in each case—hunting by Paleo-Indians then, unchecked industrialization now—remains the subject of ongoing conjecture and contention.

In place of a single season's accumulation of snow on the Kansas hills where Dodd now lives, she imagines a deep ice sheet one million years ago. In the end, we are left to ponder with her the durability of cultural memory of monumental natural disruptions tens of thousands of years ago, such as torrents unleashed as the immense glaciers of the last ice age retreated. To comprehend such events in deep time is for Dodd an act of the imagination: "What did people think, thousands of years ago, when the ice that gripped the continent I travel and call home began to melt, at first, no doubt, almost imperceptibly?"[71]

For how many generations would stories of such floods have been retold, she wonders. How would cultural memories associated with particular landscapes be kept alive, even in the wake of subsequent migration—beyond time's horizon?

NOTES

1. Dodd, *Horizon's Lens*, 25.
2. Maher, "Places Below," 57, 61.
3. Mercier, "Epic Narratives of Evolution," 253.
4. Dodd, *In the Mind's Eye*, 49.
5. Dodd, *Horizon's Lens*, 4.
6. Dodd, *In the Mind's Eye*, 273.
7. Ibid., 270.
8. Eiseley, *Star Thrower*, 27–28.
9. Lynch, "Borders Between Us," 129, 130.
10. Eiseley, *Night Country*, 234–35.
11. Dodd, *In the Mind's Eye*, 278.
12. Ibid., 280–81.
13. Ibid., 312.
14. Eiseley, *Star Thrower*, 85.
15. Dodd, *In the Mind's Eye*, 34.
16. Ibid., 34.
17. Ibid., 74.
18. Dodd, *Prospect*, 7.
19. Ibid., 95.
20. Ibid., 98.
21. Eiseley, *Star Thrower*, 56.
22. Maher, "Places Below," 62.
23. Dodd, *Prospect*, 37–38.
24. Dodd, *Horizon's Lens*, 55.
25. Ibid., 217.
26. Dodd, *Prospect*, 92.
27. Ibid., 88.
28. Dodd, *Horizon's Lens*, 204, 214.
29. Ibid., 218.

30. Dodd, *Prospect*, 42–43.
31. Ibid., 64–65.
32. Ibid., 71.
33. Eiseley, *Night Country*, 72.
34. Maher, "Places Below," 63.
35. Dodd, *Prospect*, 207.
36. Dodd, *Horizon's Lens*, 58.
37. Dodd, *In the Mind's Eye*, 99.
38. Dodd, *Horizon's Lens*, 51.
39. Dodd, *Prospect*, 82–83.
40. Pauketat, *Cahokia*.
41. Dodd, *Prospect*, 114.
42. Ibid., 112.
43. Dodd, *In the Mind's Eye*, 156.
44. Dodd, "Tsankawi," *Like Memory, Caverns*, 8.
45. Dodd, *Like Memory, Caverns*, 19–20.
46. Ibid., 20.
47. Lynch, "Borders Between Us," 129.
48. Eiseley, *Star Thrower*, 158.
49. Dodd, *In the Mind's Eye*, 221.
50. Dodd, *Horizon's Lens*, 19.
51. Dodd, *In the Mind's Eye*, 239.
52. Ibid., 243.
53. Ibid., 117.
54. Eiseley, *Star Thrower*, 84.
55. Eiseley, *Night Country*, 134.
56. See Pruvost et al., "Genotypes."
57. Eiseley, *Star Thrower*, 182.
58. Eiseley, *Immense Journey*, 46.
59. Eiseley, *Star Thrower*, 151.
60. Lopez, *Of Wolves and Men*, 96.
61. Dodd, *In the Mind's Eye*, 142.
62. Dodd, *Like Memory, Caverns*, 9.
63. Dodd, *In the Mind's Eye*, 144.
64. Eiseley, *Star Thrower*, 163.
65. Ibid., 85.
66. Dodd, *In the Mind's Eye*, 151.
67. Ibid., 34.
68. Eiseley, *Night Country*, 197.
69. Eiseley, *Star Thrower*, 21.
70. Here Dodd shares what Susan Maher terms "Eiseley's penchant for telescoping temporal and spatial dimensions" (Maher, "Places Below," 67).
71. Dodd, *Prospect*, 87.

10

Islands of Time

Paul Gruchow (1947–2004)

This is the essence of wilderness;
that is what we must not forget:
how brief life is, how unexpected,
how little of it we glimpse, how rapidly it changes.

—Paul Gruchow, *Boundary Waters*

Only grass is eternal.... I would be converted to a religion of grass.

—Louise Erdrich, "Big Grass"

Paul Gruchow's contribution as a literary naturalist lies at the cusp of ecology and memory, nature and culture. While the influence of Henry David Thoreau is pervasive, Gruchow's reverence toward the natural world and the sheer lyricism of his writing can be likened to John Muir's. He was also inspired by twentieth-century essayists such as Loren Eiseley, Rachel Carson, and E. O. Wilson, whom he credits with having bridged the humanities and sciences by synthesizing diverse disciplines.[1] Believing that the finest essays of this kind are inevitably personal and reflective, he most admired those writers informed by the scientific literature who nonetheless "democratized" the genre by addressing a wide, general audience—much as he himself sought to do. His work also follows in the tradition of American writing that explores natural history in relation to several dimensions of *time*—whether diurnal, seasonal, geological, or evolutionary. His convictions as a conservationist were ultimately grounded in an appreciation for ecological memory encompassing deep time as well as recent environmental history. By the end of his career, in fact, he had begun to articulate a more nuanced and intricate appreciation than had been advanced by previous generations of pioneering ecologists for the complex interdependencies governing succession through the coevolution of species. Above all, he believed that we must begin to stem the tide of extinction for *moral* as well as ecological reasons.

Thoreau was clearly a touchstone for Gruchow, who in his own journals reflected on the depth of this connection and expressed a profound affinity on several levels. First, there were similarities to their temperaments as fellow wanderers, prone to solitude and introspection. In fact, he regarded *Walden* as the finest book of *travel* literature an American has ever written.[2] Moreover, he relished one of its central concerns: what has been lost as our daily relationship to nature has waned—and how it might be reclaimed. Moreover, both deplored conformity and were fiercely independent. He recognized that his approach to the essay in many respects resembled Thoreau's rhetorically. Perhaps there was an element of emulation, as he would occasionally copy out passages from Thoreau's journals into his own. Above all, he sensed a spiritual kinship: "We believe, against the fashion," he declared in his journals, "in the divine content of nature... in the importance of being attached irrevocably to place...in the ultimate authority of truth to oneself."[3] Acknowledging parallels with English and German Romanticism, he shared Thoreau's penchant for contemplating philosophical and spiritual questions grounded in close observation of the natural world.

Gruchow's conception of wilderness was complex and vexing—he anticipated debates about human stewardship and management that rage to the present day. For example, while visiting Isle Royale in Lake Superior offshore of Copper Harbor at the northeastern extreme of Keweenaw Peninsula on Michigan's Upper Peninsula, he considered the role ecologists play in maintaining the native species there, particularly populations of moose and wolves. Determining appropriate conservation measures on the island had been complicated by the arrival of wolves migrating across lake ice in winter; the dwindling genetic diversity of such a small, isolated population; and severe overgrazing whenever the moose population spikes. Gruchow distilled the ecological and ethical issues at stake into three questions: "Are our wilderness preserves museums or are they laboratories? How much should they be managed and to what end? Where should the line be drawn between benign neglect and the failure of stewardship?"[4] These in turn raise the more nuanced issues of degree, strategy, and ethical reckoning. Just when and how, he wondered, should we attempt to steer natural systems toward some idealized or preexisting state? Moreover, what purposes are to be served by protecting significant tracts of land where ecological integrity remains largely intact? He acknowledged the conventional aim of preserving biological resources that might fulfill utilitarian ends, such as a repository of botanical sources for formulating medicines. He deemed such rationales self-evident. In the long run, species diversity might also provide an ecological safety net, a library of genetic information to develop new crops, or to fortify existing ones. Moreover, he already sensed by the 1980s that the ecological consequences of human alterations to the planet were approaching the absolute. "Preservation of the biosphere in all its present richness and diversity is not merely expedient," he declared at a meeting of environmentalists in 1988, but "essential to the very survival of life itself."[5]

Gruchow understood that the conventional thinking of ecologists and wildlife biologists at the middle of the twentieth century had begun a subtle but fundamental shift away from models based on succession and toward a presumably static climax community. The new thinking, which he fully embraced, posited ecological change as a constant process. Above all, he grasped that preservation was being redefined in bold new ways. In "The Kingdom of Grass," an unpublished manuscript completed in 1995, he offered his most compelling synthesis of new directions in conservation emerging at that time. E. O Wilson's theory of island biogeography had recently been applied in other geographic contexts, and Gruchow knew that historically indigenous species were unlikely to survive let alone thrive on undersized and isolated parcels. For that reason, he advocated a system of preserves linked in strategic ways, such as establishing corridors for migration, which has since become a common conservation practice. Moreover, echoing bioregionalists like Gary Snyder, he argued that, ideally, entire watersheds must be protected. Much like iconic figures advocating restoration, from Gene Stratton-Porter to Wendell Berry, he maintained that "a very poor remnant of prairie—one that has been used as pasture for many years but never plowed, for example—can sometimes be nursed back to health."[6] And like Aldo Leopold, he urged rural landholders to reintroduce native plants essential for regenerating habitat.[7] He believed that natural history, beyond addressing the workings of biological processes per se, must attend to cultural significance—a much more inclusive view, one that complements ecological knowledge with the emotional, ethical, and spiritual dimensions of our relationship to the rest of life on earth. Indeed, he believed that "natural history inevitably includes a sense of oneself in relation to what is observed."[8] Such synthesis matters, he realized, "because we share with it a common past, and because, once we become aware of that commonality, we know in a deeper and truer way who we are."[9]

Like preceding midwestern naturalists, Gruchow studied the landscape for evidence of change and recognized the cumulative impact of the ongoing expansion of cropland. He understood how the erasure of the grasslands and wetland ecosystems in the Upper Midwest had profound environmental consequences: "As we have drained the swamps and marshes, leveled the forests, farmed the prairies, and diverted the waters of the western rivers for the greening of the deserts, we have done something more pervasive than simply destroying a multitude of local habitats.... We undermined the framework for biological diversity."[10] And he was alarmed by the extent of this transformation: "Only about one-tenth of one percent of the native tallgrass prairie that covered the central United States at the time of white settlement remains."[11] His prognosis for the biome as a whole was severe: "So much of the original prairie has been destroyed that even were we to preserve every remaining remnant, we still might not be able to save many prairie species."[12] The toll on wildlife of the region had undeniably been extreme, yet during his youth the region's verdant landscapes left an indelible mark on his imagination, and instilled a lifelong dedication to contesting relentless development that diminished their ecological integrity. "The prairie itself, as an ecosystem, is dead too," he declared, since "most of its wetlands have been drained, and many of its rarest species have become extinct. There are only a few places left in the United States where you can see the prairies as they looked even a hundred years ago. The lessons that the prairies might teach us," he lamented, "are almost lost."[13]

In *Worlds Within a World*, Gruchow's depiction of the Ripley Esker reserve, one of a dozen designated ecological preserves in Minnesota, he suggests several centuries of landscape change still unfolding: "I saw the remnant bur oak savanna, a prairie landscape, and beyond it an old field reverting to savanna," with the use of controlled burns to favor native species.[14] This capsule narrative of environmental history encompasses three distinct phases: native grasslands altered by settlement and cultivation, reversion to a feral state once farmsteads had been abandoned, followed by subsequent management and regeneration. This progression parallels patterns of environmental history in other regions such as New England, as chronicled by William Cronon and Tom Wessels. While a scarcity of trees on the plains may have delayed development by white pioneers for generations, eventually successful settlers embraced technologies that would lay bare its rich soils for conversion to cropland. Inventions such as windmills and barbed wire in the 1870s, however, soon encouraged overgrazing across vast tracts of the prairie, much as the steel plow had begun the era of sodbusting a generation earlier. Taken together, Gruchow declares, "in scarcely fifty years, the American grasslands had vanished and a way of life that had prevailed on the landscape for tens of thousands of years was at an end."[15] What remained on the once expansive prairie was a veritable desert of cornfields, vast tracts of monocultural croplands plowed "to its last square inch."

As Scott Russell Sanders concludes, "By the mid-twentieth century, when Paul Gruchow was growing up in southwestern Minnesota, the tallgrass prairie had become little more than a memory. All but a few remnants of these millions of acres had been plowed . . . vast herds and flocks had dwindled or entirely disappeared"; nonetheless, "the natural forces that had created the tallgrass prairie were still in play."[16] Like other naturalists in the Midwest, including Aldo Leopold and Paul Errington, Gruchow advocated preserving representative examples of a variety of naturally occurring ecosystems for posterity. He contemplated the fate of remaining fragments of wetlands on the grasslands and understood that by functioning as refugia, such wilderness preserves comprise biological islands in time. Places like Cayler Prairie in northwestern Iowa seemed emblematic to him of midwestern environmental history: a microcosm of the once vast tracts of grassland biome that have been plowed and planted with a handful of crops, predominantly corn. He realized that such a preserve represents an ecological oasis, since so much of the surrounding landscape had been converted to cropland: an "unusually large prairie remnant for cornbelt country, 160 acres . . . tallgrass prairie landscape as it might have been a couple of centuries ago."[17]

A generation earlier, the region had become a crucible for early ecological thought in the wake of agricultural practices that had profoundly altered the tallgrass prairie. Raised on a small farm in the Upper Midwest, Gruchow grew up on the frontline of upheaval sweeping the region as agribusiness displaced millions of families—including his own—twentieth-century farmers who tended a variety of traditional crops and livestock on relatively modest plots, a model of self-sufficiency rooted in the

Jeffersonian tradition.[18] His grandparents, who farmed as homesteaders, knew the region's hardships well, having buried their first eight children there.[19] He described his own family's subsistence lifestyle in the 1950s on a seven-acre truck farm. Following the seasonal round of the almanac, they tended poultry for meat and eggs, goats for milk, hives for honey, and wheat for flour, as well as a cornucopia of vegetables for the table, supplementing their diet with wild game. By diversifying crops and livestock, he explained, they were bucking the tide of monocultural operations overtaking the landscape: "big open fields of corn and soybeans stretching to the horizon."[20] Many such family farms, historically the mainstay of rural communities throughout the region, became increasingly less viable financially, marginalized in the new economy.

As a child he witnessed the advent of modern, large-scale farming methods at a time when the social and environmental consequences of this new agriculture were only beginning to be understood. He was born the very year that "the first miracle chemical of industrial agriculture was introduced" in 1947—a moment when the rhetoric of progress and efficiency was widely marshaled to justify a fundamental reorientation of American agricultural practices.[21] That industrial model, he argued, has largely eclipsed the deep and meaningful appreciation of nature gained by the direct personal engagement entailed in tilling and harvest. He lamented both the human and ecological costs: "In the decade of my coming of age, millions of farm dwellers left the land.... If you grew up on a farm in the last fifty years, as I did...you would have noticed the diminishing songbirds, the disappearing butterflies, the vanishing potholes, the uprooted fencerows, the balding hilltops."[22]

Returning years later to the site of his family's farm in Rosewood Township, he reported that fencerows that had once provided forage and shelter for a host of small mammals and migratory songbirds were no longer to be found. Even the marsh was gone, with alarming consequences for the very wildlife that had inspired his interest in natural history: "The waterfowl are gone, the raptors are gone, the burrowing animals are gone, the predators and herbivores are gone."[23] And once gone, he realized, such thriving wild communities are all too soon forgotten: "The heritage has so thoroughly vanished, in fact, that it no longer survives even in memory."[24] The great herbivores have long since disappeared, along with "all that once was and might still have been,...the prairie chickens that no longer boom,...the pelicans that roost elsewhere,...the wolves that have ceased to howl in the night, ... the elk and pronghorns that have vanished into the west."[25] With the antelope, elk, and bison gone, what remained, he concluded, was indeed "a pale ghost of the world that once existed in this place."[26] Moreover, he questioned the premises of agricultural practices that completely displaced wildlife, particularly how the region's pothole lakes had been drained by the tens of thousands; these were wetlands that had provided breeding grounds for flocks of waterfowl numbering in the millions. Populations of game birds such as grouse plummeted as their native habitats vanished. Species previously subjected to overhunting or eradication as pests were among the most vulnerable locally, while some were in danger of outright extinction. His accounts of lost abundance resemble those of Errington, who had chronicled similar declines in wildlife population in Iowa and the Dakotas a generation earlier. Indeed, the narratives of environmental history by the region's pioneering ecologists are a repository of cultural memory, chronicling landscape change as well as reflecting our shifting relationship to nature over time.

Gruchow viewed such trends as both ecologically "brittle" and socially destabilizing, causing "the impoverishment that befell every aspect of rural culture with the industrialization of farming"; agribusiness was "the death of the culture of agriculture."[27] Native ecosystems were often greatly compromised, with many uprooted or undermined entirely. Much of the surface water essential to indigenous flora and fauna had been drained away, and half the fertile topsoil was lost to erosion in a single century, leading him to lament that "the richest and most representative prairies—those on level black-loam soils that yielded the richest agricultural region on earth—have all but disappeared."[28] This momentous transformation of the landscape precipitated a decline of rural communities throughout the region. "We have, over the past half century," he concluded, "driven all but a handful of our farmers from the land, at an enormous cost to soil, water, fuel, and biological diversity."[29] Moreover, he recognized that agricultural practices predicated on increasing dependence on petrochemicals and mechanization were unsustainable environmentally. While conversion to cropland may have proceeded incrementally, the cumulative impact has been monumental ecologically, especially on the diversity of native species: "We have steadily labored at turning one of the most fertile places on earth into a desert," he asserted, adding that lest we mistake such a claim for hyperbole, "a square mile of conventionally cropped corn is biologically poorer than a square mile of arctic tundra."[30]

A litany of species had been lost: "The prairie world is now crisscrossed by a labyrinth of tiles and drainage ditches.... With the disappearing ponds have vanished, many of the waterfowl, many of the insects, many of the predators."[31] Recounting an excursion to Otter Creek with friends one Christmas, he recalled how developers in the nineteenth century had drained the marshes by channelizing waterways with steam-powered dredges, leaving a "naked ditch" devoid of wildlife to this day and making its name a cruel irony: "nowhere any sign of life or habitation, no bird overhead, no nest or burrow underfoot, no track in the mud, nothing to give credence to the name Otter Creek.... A romantic reference to time past."[32] As he offered a brief prayer of repentance, the naturalist in him held out hope for redemption: restoration through the biological imperative of ecological memory, with its potential for regeneration of historically indigenous species.

Gruchow's narratives of environmental history take on an elegiac tone when he contemplates the mounting ecological costs. Well aware of the extent of what had already been lost, he championed the preservation of the native ecosystems that remained and advocated that rural students become acquainted with "the ecological and geological history of the place."[33] A single site such as Minnesota's Blue Mounds, for example, can provide habitat for thousands of species, including at least 220 birds and 30 mammals, as well as a smattering of reptiles and amphibians such as snakes, turtles, and salamanders, giving us a glimpse today of the historical biodiversity of the region. As ecocritic Kendra Smith observes, "Paul Gruchow contend[s] that knowing one's place also engenders a sense of belonging and responsibility because it enables a person to understand one's dependence on the natural landscape and on other members of shared communities... [and] as people have fewer opportunities to directly observe nature and to engage with one another, they know less and less about the natural and cultural history of the places they live."[34] In his journals, he lamented that traditional bonds to place stretching back across so many generations had been largely severed in transitory modern societies.

As a boy, Gruchow had been enraptured by unsettled places he perceived to be wild, and for a time he even fancied becoming a backcountry trapper, much as Errington had done, romanticizing the rugged, primitive lifestyles glorified by adventure and sporting literature. He imagined mastery of traditional woodcrafts, including trapping, as a way of engaging the wildlife at an elemental level. To hunt or trap successfully, he discovered, necessitates acquiring an intimate acquaintance with animal behavior as well as learning to read the ecological intricacies of a landscape. Reflecting on such childhood experiences later, he realized that pursuits that are likely to strike us as anachronistic today such as trapping had commonly been practiced by rural youth throughout the Upper Midwest well into the twentieth century. He believed the underlying impulse to be an archetypal one, still widely shared: "to undertake shooting or trapping as a rite of passage… an uncanny act of cultural recapitulation: The age at which boys become trappers is precisely the age at which, in pre-agricultural societies, boys left behind childish pursuits to take up the adult labor of hunting."[35]

"I was after a share of wildness in the creatures I was pursuing," he wrote. "I wanted to be able to read the landscape…to assume its language…. I yearned for that leap of imagination that would send me off into the unimaginable wildness."[36] His abiding fascination with the natural world, he realized, had been catalyzed by those youthful adventures in the outdoors, echoing a refrain commonly found in memoirs by field biologists of the time. Formative childhood experiences, he points out in "The Meaning of Natural History," represent a "convergence" in the biographies "of most important naturalists in American literature."[37] Moreover, such early experiences afield can inculcate a heightened empathy and lifelong appreciation for wildlife and, as others from Thoreau to Errington have observed, engender a sense of environmental ethics. He believed that what had begun as a youthful lust for hunting could mature, as it had for him, into reverence and a sense of responsibility.

Throughout his life, Gruchow retained a capacity for wonder afield and a keen desire to engage the natural world in direct and primal ways. He described the calling of literary naturalists in precisely this light: "They are all, to the last one, celebrants of nature."[38] In his second book, *Grass Roots*, he recounted early encounters with open land neighboring his family's pastures, a landscape he characterized as "remnant virgin prairie."[39] The place that initially captivated him there was actually a bog—his first glimpse of the ancient wetland ecosystems of the tallgrass prairie—verdant and teeming with life. A cattail marsh in particular inspired profound reverence. He recognized the staggering concentration of life in such places and intuitively grasped that even smaller pothole lakes were wellsprings of species biodiversity. Moreover, he relished their antiquity—having "existed for thousands of years, a surviving testament to the tallgrass prairie" with its miles of roots and unseen rivers underground.[40] Astonished by sheer biological abundance, he marveled how such ecosystems centered on surface water supported "as rich a diversity of life as anywhere on the prairies."[41]

> Frogs, snakes, and insects numbering in the thousands of species; and so came the skunks and weasels, the minks and foxes, the raccoons, the mice and shrews, and on the uplands the burrowing animals, the ground squirrels and badgers, and pocket gophers; and in pursuit of them came the wolves and coyotes, the raptors, the hawks and owls; and in the tall grasses around them grazed the great herbivores, the bison, elk, and antelope... and the water lasted from year to year, attracting muskrats, and turtles, clams and crayfishes.[42]

This litany of species once native to Minnesota is reminiscent of Errington's accounts of the abundance a generation earlier in the Dakotas at the time of settlement, when extensive wetlands still provided habitat for a wide variety of native wildlife to thrive. Gruchow recounted how the prairie biome had historically been anchored by nearly one hundred varieties of grasses, each species a marvel unto itself: "the big bluestem, taller than a man, and, in September, the color of fine wine; the needle sharp grasses with their twisted seed pods, sharp as pins, that wind themselves into the earth like corkscrews; the bearded plumes of the Canada wild rye; the feathery spires of the Indian grass; the little tufts of blue grama grass with flowers."[43] To his eyes, the profusion of hundreds upon hundreds of types of wildflowers elevated the grasslands to a veritable garden, from the first phlox of spring to the profusion of plants blooming in autumn. Beneath them a labyrinth of roots held the fertile soil against the erosive forces of wind and water. He treated such themes lyrically in poems, such as when describing bluestem in "Reasons for Living 9":

> its flowers and blades conspiring
> to conceal the hundreds of miles
> of vital roots in fertile darkness.
>
> A prairie... is
> like a forest whose canopy grows,
> underground.[44]

Not until the catastrophic drought that triggered the Dust Bowl would the profound ecological significance of prairie sod be fully understood.

Beneath this protective turf, a host of smaller mammals burrowed, including pocket gophers and ground squirrels, badgers and prairie dogs. Yet the abundance and diversity of such diminutive creatures, he realized, is all too easily overlooked. Bird species were particularly numerous, including over one hundred kinds of songbirds, as well as raptors, owls, and millions upon millions of migratory waterfowl. Moreover, countless pothole lakes were once populated by multitudes of amphibians, particularly frogs, as well as reptiles such as turtles and snakes. In the long run, he reckoned, ecological complexity ensured vitality: prairies composed of an enduring mosaic of thousands of separate species that had coevolved over millennia. Indeed, like Leopold, he saw such biodiversity as evidence of ecological integrity and a measure of its resilience. "One of the principles upon which the prairie was made," he wrote, was "the principle of diversity. It is in an alliance of differences that the prairie finds its vigor; it fashioned its resilience and stability."[45]

By contrast, monocultural cultivation, he understood, is ultimately unsustainable—even precarious—due to its dependence on a handful of plants. In "Corn Is Not Eternal," he described how only a century or so before, many Native American peoples had come to rely heavily on the buffalo that then numbered some sixty million in North America. "We have made upon Indian lands and in place of the prairies a new kind of grassland based upon corn rather than upon big bluestem or buffalo grass," he warned, "and that culture depends on corn as completely as the Lakota depended upon the buffalo."[46] Similarly, during the twentieth century, he reasoned, coming to rely increasingly on a single crop such as corn as the basis for so many processed foods was to court disaster, since "it would take exactly one persistent pathogen to devastate our culture as we know it," not unlike the Plains Indian cultures undermined by the precipitous decline of the bison.[47]

Who could have imagined how vulnerable the once seemingly inexhaustible herds would become—or how suddenly they would nearly disappear? Gruchow lamented how the species had long since been eradicated across much of the continent: "It has been a century since buffalo wandered freely in that country, but their ghosts linger in the grass," as well as in our memory and imagination.[48] At moments that memory seemed palpable to him: A haunting passage in the closing chapter of *Grass Roots* describes listening in the dark to the howls of dogs on farms in the distance: "I could hear in their voices the ancient cries of gray wolves in the days when great herds of bison roamed the plains and moonlight dances in the endless waves of grass. I could feel then the wilderness in my own bones."[49] When a member of his party unearthed a buffalo skull from an embankment while visiting a stretch of the Oregon Trail in Nebraska, he described the impressive visage of this creature with its "wickedly pointed horns, the breadth of its hairy brow, the wide spacing of its enormous black eyes." "A buffalo" he concluded, "looks ancient."[50] Later, observing a remnant herd preserved at Minnesota's Blue Mounds grazing the pasture in a fenced enclosure, he was again reminded of the region's environmental legacy in deep time. "These survivors," he declared, "are the ghosts of the lost world of the prairie."[51]

Gruchow's first book offers a comprehensive account of the cultural significance of the buffalo to those Native Americans for whom they were the very staff of life; he exhaustively catalogs their uses for its hides and horns, sinew and hooves, organs and bones. Indispensable, their continued abundance had long been a cultural article of faith: "For thousands of years it seemed as if there could be no end to the bison. The Plains Indians believed they were eternal."[52] Yet their eradication was unmercifully swift: "the last of the southern herds of bison expired" in 1874, he reports, following a three-year campaign that had slaughtered over four million animals, turning the prairie into a boneyard. Farther north, "farmers settling in Nebraska and Kansas in the 1870s sometimes found it necessary to pick the bones from their claims before the land was fit to plow."[53]

Gruchow sympathized with the plight of the Plains Indian, deprived of game that had sustained an entire way of life, particularly following introduction by Europeans of horses in the Americas. For once the great herds had been extirpated, such traditional lifeways would become untenable. "As the buffalo died," he explains, "so died whole nations and cultures on the Great Plains," as prophesized by the Pawnee who believed "that the father of all buffalo lived in the north at the junction of heaven and earth and that every year he shed a little hair. When the last hair had been shed, they said, life would end."[54]

Yet what fascinated Gruchow most was the complex role of this formidable animal in Plains Indian life; besides providing sustenance and the makings of material culture, it had taken on great symbolic and mythological significance. In his journals, Gruchow began to conceptualize plans for an ambitious book about the American bison—one he did not live to complete—coupling natural and cultural history that spanned the centuries from first European contact to the tragedy at Wounded Knee. He hoped that such a study could ultimately encompass indigenous cosmologies based on the spiritual dimensions of a sacred relationship to nature.[55]

Indeed, the Mandan, as well as many other Native Americans reliant on the bison, he recognized, forged one of the closest relationships between human beings and a single animal the world has ever known—one encompassing religion.[56] He described how Black Elk had witnessed their disappearance, which inspired the Lakota holy man's vision of a cultural renewal. In fact, he observed, the Ghost Dance religion was predicated on the reappearance of the vanquished herds from the underworld. Sun Dance ceremonies similarly sought to ensure their enduring strength and return year after year. Such rituals were an expression of reverence and gratitude that maintained a sacred balance between a people and the creatures on which their survival depended. Yet such heavy dependence on a single species, he recognized, left Plains tribes vulnerable.

Like other literary naturalists before him and since, Gruchow contemplated the complex cultural legacy of the Americas. He challenged conventional stereotypes that posited Indian identity in essentialized ways, such as monolithic depictions of nomadic buffalo hunters of the Plains, stressing instead the linguistic and cultural diversity to be found historically among hundreds of tribes across the continent. He pointed out how the Mandan practiced an elaborate array of subsistence activities, supplementing wild edibles and cultivated crops with seasonal rounds of fishing and hunting—above all, buffalo. Understanding how a number of Native American nations engaged in shaping the grasslands by regularly burning it to create habitat attractive to game, he characterized Plains Indians as "agents" of fire, and observed how burning had also been employed historically in southeastern Minnesota at Mound Prairie.[57] Like naturally occurring conflagrations, he realized, these blazes played an essential role in the dynamic equilibrium of the prairie biome.[58] He grasped how such subsistence practices are tethered to seasonal cycles and so connected to specific places—and understood as well the potential for religious expression of such a relationship.

These reflections were often occasioned by traveling to places where Native Americans had left major traces on the landscape. Like Scott Russell Sanders and Elizabeth Dodd, who have also pondered the Ojibwe pictographs painted on islands in Minnesota's Boundary Waters, Gruchow sensed how such ancient art can stir our imaginations today in powerful ways. Some of these figures are thought to have survived up to two thousand years, and while the sources of pigments are known—such as iron, eggshells, and sturgeon—many compelling questions remain unanswered:

"Why they were made; or why they were made *where* they were made; or what, in general, they signify; or to whom they were addressed."[59] No doubt those images derived from mythic and historical cycles of stories had symbolic resonance for their makers. In "The Grace of the Wild," originally published as *Travels in Canoe Country* and reprinted in *Boundary Waters*, Gruchow acknowledged that little is now known about the peoples who created the paintings, let alone their "intentions." "The battle of the human heart," he quotes Loren Eiseley as saying, "is endlessly waged between memory and forgetfulness."[60] Some of the figures seem representational: "A few of the markings are immediately recognizable: a canoe, a moose, a human figure, a thunderbird," he wrote, while others remain "utterly mysterious."[61] Were such images connected with rituals to ensure success of the hunt, he wondered, or perhaps linked to shamanistic rites? It is tempting to speculate, for example, that depictions of animals favored as game correspond to religious beliefs and practices associated with the hunt. Since successful hunting was essential to a tribe's survival, ceremonies expressing reverence carried profound—even *existential*—significance.

Pondering the meaning of ancient art and its possible purpose as prayer, Gruchow approached such sacred sites with humility. He realized that interpreting what they were once meant to represent—let alone *why* they were originally made—remains subject to conjecture. He acknowledged the difficulty of interpreting the significance of the glyphs fully since cultural memory had been disrupted, including the narratives of their creation, meaning, and function. Since the symbolism of geometric figures such as spirals and concentric circles continues to elude us, the mystery evokes wonder while tantalizing the imagination. Indigenous peoples, he concluded, "came to know the place spiritually.... This is what it means to be native to a place: to know it intimately enough so that one can say where lives its spirit, or spirits."[62] Ultimately, he confided, "I hope that they might tell me something about what it means to be indigenous."[63]

Visiting Medicine Wheel monument in the Big Horn Mountains of Wyoming, Gruchow contemplated another historical trace left by Native Americans on the landscape. He described that journey and the profound resonance he experienced there in "Medicine Mountain," an essay critics have hailed as the most incisive writing in *The Necessity of Empty Places*. Medicine Wheel struck him as hallowed ground, an impression amplified by the grandeur of its panoramic vistas of surrounding country and distant horizons. The site is still in ceremonial use, he reports: "To this day the ancient rites of the native Americans are celebrated there, and...worship of some kind has been practiced there for at least a couple of centuries."[64] Similarly, brightly colored prayer bundles left by today's Indian pilgrims are still to be found at Bear Butte (also in the Big Horn Range), traditionally venerated for vision quests by Plains tribes, and a place familiar to Crazy Horse.[65] From its summit, the Black Hills sacred to the Lakota are visible to the south.

Since the Medicine Wheel became known to Europeans in the 1880s, ethnographers and astronomers have speculated about its antiquity, purposes, and symbolism. Its possible use for astronomical observances, for instance, is suggested by the precise placement of cairns that align with solstice points, or presumably had done so in the past. The overall configuration of stones at Medicine Wheel has also been likened to a sacred geometry found in the design of earth lodges of the Pawnee, the tipis of the Cheyenne, and above all the medicine lodges associated with the Ghost Dance. Twenty-eight spokes radiate from the circle's center, corresponding to days of the lunar month, ribs of the buffalo, and arithmetically the product of seven and four, both sacred numbers in Native American cosmologies several times over. Yet Gruchow approached such conjecture with caution, since such theories are necessarily speculative once the cultural history has been forgotten. The Crow Indian Nation that today inhabits the area adjacent to the site knows only that it predates their arrival in the area. We do not know how the monument came into being, he concluded, or why. These mysteries still beckon us, often evoking a sense of reverence. After its makers moved on, the narratives that once animated this sacred place fell silent. Consequently, modern interpretations remain provisional, since we are reading the indigenous text of this landscape but lack the cultural memory that brought it into being.

In a similar way, reading natural landscapes for narratives of *environmental* as opposed to cultural history also entails imagining what had come before, such as recognizing the way ferns and fungi, moss and lichen of a northern forest floor harken back to an epoch before flowering plants had even evolved. Gruchow recounts a succession of ecosystems in the wake of glaciation, as tundra studded with spruce were followed by bur oak savanna, and finally the grasslands whose extent bewildered the first Europeans to explore them. After the last ice age, grasses that had evolved perhaps eighty million years ago began to reshape and ultimately dominate the prairie landscape. Nomadic hunter-gatherers moved their seasonal encampments in anticipation of the migration of game and according to the cycle of ripening and harvest; yet by the time of European exploration, the vast majority of Native American peoples across the continent were also cultivating corn in virtually every region where the climate and growing season allowed. Consequently, subsistence skills passed from generation to generation included a place-specific mix of hunting and gathering practices often in conjunction with the cultivation of crops such as beans, melons, squash, sunflowers, and tobacco. When in 1541 Coronado crossed the plains of what today is Kansas, he reportedly likened the abundance of plums, nuts, grapes, and mulberries to those grown in Spain.[66] The sustainability of such pre-contact agriculture, Gruchow reasoned, was not due to ecological reckoning but rather mobility: once a tract of arable land was worn out and lost its fertility, villages were simply moved and new fields cleared.

By contrast, when European settlers arrived in the Upper Midwest, the grasslands adjoined woodlands that spanned over thirty-four hundred square miles of Indiana, Iowa, Minnesota, and Wisconsin. A single square mile of the "Big Woods," considered the most extensive of remnants remaining today, is an island ecosystem that still supports a number of especially primitive life forms. These include mosses, ferns, and horsetails, as well as some of the earth's first flowering plants and species such as the dwarf trout lily, which is "among the rarest plants in the world" and occurs only in a handful of surviving fragments of Minnesota's original Big Woods.[67] Gruchow's depiction of such refugia underscores how surprisingly dynamic and precarious such places can be: "Though ornamented with plants of venerable lineage," he explains, "the Big Woods is only about three hundred years old. When the climate suddenly became warmer and wetter," today's maple-basswood forest replaced the oaks that had once dominated.

Because the grasslands repeatedly advanced and retreated, Gruchow realized, its boundaries were transitory: "Where trees and grasses have battled for supremacy, fire has been a critical influence. When fires prevailed, grasses extended their dominion. When they did not, trees took hold."[68] Moreover, he recognized just how dramatic the effects of such gradual ecological processes could be: "The lands I have come to thinking of as naturally prairie," he realized, "were, in fact, forests."[69] The idea of such a dynamic successional pendulum in deep time had intrigued Leopold as well. Tenacious bur oak pioneered drier terrain by surviving fire, while cottonwood, chokecherry, and willow might colonize riparian corridors.

Gruchow recognized that the dynamics of the biome as a whole had prevailed for many millennia: colonization and succession showing resilience following flash floods and prolonged drought, severe weather and infernal fire. At Cayler Prairie in northwest Iowa, for example, he readily recognized the geological processes that had once shaped many midwestern landscapes, regenerating after ice retreated. Situated on a terminal moraine, it reveals to this day traces of ancient glaciation.

Among the creatures that inhabit the region's wetlands, including kettle lakes shaped by glaciers, are several ancient species such as the snapping turtle, which evolved close to three hundred million years ago. Similarly, the dragonfly is part of the legacy of a primordial ecological order and has found a way to survive for hundreds of millions of years. Gruchow reminds us that the dragonfly was among the very first flying creatures on earth. He grasped that the biological imperative of ecological memory was a key mechanism ensuring the survival of such native species—the dynamic capacity for regeneration following disruptions, which enabled aquatic ecosystems to persist over the ages. Seeds dormant for many years remained viable, while grasses were especially resilient: well-established systems of deep roots could be revived after decades, whenever favored by ample rain.[70] He described how cattails, the plant that plays a pivotal role in the latter stages of the dragonfly's metamorphosis, can be rekindled from seeds that have been long dormant: a marsh could spring forth from a place that had been parched for a quarter of a century.

Great antiquity also heightened Gruchow's sense of reverence toward species such as sandhill cranes, among the oldest of all birds. Fossils of this archaic bird date from at least fifty-five million years ago. They are a vestige of the age of dinosaurs, from which all birds have descended. The species seemed to him a living relic that has managed to survive eons of environmental change. The cry of the cranes also awakens connotations of great migrations predicated on ecological memory. Some routes of up to six thousand miles each way span nearly half the circumference of the globe, and rely on particular types of habitat across the hemispheres to provide forage and suitable breeding and wintering grounds. Whatever had originally prompted sandhill cranes to undertake such an astonishing journey in the first place, the spectacle of their arrival on the Platte River in March and April inspires awe. Gruchow contemplated how routes from such far-flung places as Tierra del Fuego near the tip of South America effectively tie the Upper Midwest to the planet as a whole—"bearing news of distant worlds," he mused, and reminding us "that the prairie world is not a world unto itself. It is a world within a world."[71] The sight of such an epic migration inspired a lyrical flight of the imagination: "Part of me ached to follow them, to rise up on wings of my own, to fly to some wild and unbounded place."[72]

Especially given the evolutionary history of such species in deep time, their continuance has profound significance. Gruchow returned every March to the Platte River in Nebraska, a personal pilgrimage to witness once more a migration rivaling any in the world. In 1987, for instance, he described in his journals visiting for the tenth time in three years, likening the experience to dream or prayer, entranced by the "hypnotic chant of the cranes."[73] He marveled how the birds return each year, monumental flocks synchronized to seasonal changes in light. Observing their stylized rite of "dancing" in greeting and courtship—along with their especially distinctive call—prompted him to contemplate the ongoing survival of such seemingly primeval species. "We need to be reassured that there is order and logic in things," he declares in *Grass Roots*, "that there is predictability and regularity in life, that some things stand fast and hold firm. The cranes never disappoint us. The cranes are always there."[74] For Gruchow, as Loren Gustafson concludes, the sandhill cranes represent nothing less than "a reassurance of order behind change.... Reassurance comes from the collective migration of cranes... [confirming that] their collective journey will continue unchanged."[75]

Gruchow's writings about backcountry experiences in rugged terrain of the Rocky Mountain West were often framed by recollections of the Midwest. Wherever he travelled, in fact, he consistently brought to bear an appreciation for environmental history forged in the region. "In all these journeys," Lou Martinelli observes, "the damaged tallgrass prairie and Gruchow's agricultural roots will not be forgotten."[76] Hiking the remote alpine foothills of the Big Horn Mountains, for instance, his thoughts returned to the extirpation of many species throughout the Midwest. He was reminded that today's inhabitants of the grasslands no longer recognize how dramatically the region's wildlife has diminished: "In my prairie town, the upland sandpipers and the bobolinks are mostly gone and the meadowlarks become rarer every year.... White pelicans no longer nest there, but wood ducks do, in artificial nests. The gray wolf is extinct.... The otter has disappeared."[77] In their place are the coyotes, opossums, and ubiquitous white-tailed deer—species adept at colonizing the outskirts of factory farms. "Most prairie children do not now encounter any of the magnificent large mammals of their place," he continues, "the grizzlies, the elk, the antelopes, the bison until they take a trip to one of the sanctuaries in the intermountain west."[78]

If the early books, such as *Journal of a Prairie Year* and *Grass Roots*, attempt "the reconstruction, through memory, of our damaged prairie ecology," as Martinelli contends, subsequent essays are set in a variety of unsettled places that still represent "ecology that largely survives, but is vulnerable":[79] whether Isle Royale and the Quetico-Superior Wilderness described in *Boundary Waters*, Minnesota's designated scientific and natural reserves portrayed in *Worlds Within a World*, or remote landscapes of the Rocky Mountain West explored in *The Necessity of Empty Places*. Such journeys became pilgrimages, inspiring spiritual overtones and reflections on our perception of nature as sacred. His ambitious essay "The Grace of the Wild," reprinted in *Boundary Waters*, is formed of nine sections mirroring the canonical hours of the Roman Catholic Church and has been likened to Biblical psalms.[80] Indeed, Gruchow believed that "the beginning of spirituality" could come from encounters with such places: "Wildness matters not because it alone is sacred," he reasons, "but because it arouses in us the sense of sanctity that makes visible the sacredness of everything else in life."[81] He concludes: "the way to restore our sense of interconnectedness of all life is to rebuild our memories of nature by experience."[82] He came to believe that our sense of place is reconstructed through ecological memory and that therefore the study of natural history must encompass our own ever-changing relationship to the rest of life on earth.

Rather than hewing to any Romantic notion of an idealized, "pristine" wilderness, Gruchow grappled with the deeper ethical dilemma of what forms of stewardship are still possible given the deteriorating state of the natural world—and what purposes they might ultimately serve. The sad irony, he observes, is that "we are now condemned to be nursemaids to a natural world that is ailing because of our improvements upon it."[83] This is "the paradox of wilderness," he declares. "It was only when we had already lost it that we could begin to see the value."[84] He laments that it is "still unclear how to reinsert ourselves into nature, which our survival requires, given the biological constraints of life itself."[85] Even preserves and refugia protected from development, he realized, become increasingly vulnerable and volatile precisely because they have become progressively more fragmented and isolated. In the context of environmental history, such surviving islands in a sea of development (such as remnants of tallgrass prairie surrounded by seemingly interminable tracts of cropland) represent an irreplaceable ecological legacy.

He held no illusions about human stewardship, however, recognizing that returning nature to any venerated ancient state was already beyond reach. This in turn led to a rethinking of the purpose of the preservation of a wide variety of ecosystems where processes of disruption and renewal were ongoing. "A patch of prairie is not a museum piece," he declares, "but a living experiment, a laboratory of constantly evolving ideas ... a changing world."[86] Such places, he realized, were each a unique repository offering alternative strategies for biological survival of the biome, which humans might ultimately learn to heed. "When we have exhausted our soils and need to learn how to make new ones," he wonders, "where will we go for instruction? Into the wilderness," he concludes, "where this delicate and complicated work of reconstruction is constantly in practice."[87] As early as the beginning of the 1990s he had publicly championed a bold alternative to modern agriculture: a "newly emerging paradigm is that of the ecological economy which values diversity [and] sustainable yields."[88] He envisioned an integrated approach that emphasized diversification of crops tailored to particular places—patterned in part on preexisting native ecosystems. Yet, he conceded, even full-fledged "prairie restorations are probably, at best, only approximations of the real thing." Indeed, he maintained that if any hope remains at all, "Our damaged ecologies—the places we live—can be made whole in part by remembering them when they were whole, then living in such a way as to restore them."[89] He believed such attempts at conservation were likely to be beneficial in significant and unanticipated ways. "The very difficulty of creating even a restored prairie teaches us something valuable about the beauty and complexity of the prairie world."[90]

NOTES

1. Gruchow was an exceptionally well-read and attentive naturalist, and his narrative and reflective writings frequently cite Thoreau, as well as other major American naturalists and conservationists. His essay "The Meaning of Natural History" discusses Rachel Carson, Aldo Leopold, John Muir, and E. O. Wilson.

2. Pronouncing *Walden* a singular work, Gruchow conceded that he also admired a variety of modern essayists—many with a proclivity for natural history—including Edward Abbey, Diane Ackerman, Wendell Berry, Bruce Chatwin, Annie Dillard, Sue Hubbell, Peter Matthiessen, John McPhee, Richard K. Nelson, Gary Snyder, and Paul Theroux.

3. Journals of Paul Gruchow, February 18, 1987. Permission to quote archival material granted by Louis Martinelli, director of the Paul Gruchow Foundation. All archival material is located in the Paul Gruchow Papers in the Southwest Collection/Special Collections Library, Texas Tech University, Lubbock, Texas.

4. Gruchow, *Boundary Waters*, 188.

5. Gruchow, *Uses of Wildness*, 9–10.

6. Gruchow, “Kingdom of Grass,” 103.

7. Promoting “land health” and conservation generally, Leopold exhorted rural citizens throughout the region to adopt practices advocated by professional biologists on public lands, such as setting aside acres of prime habitat in close proximity to their cropland where natural ecosystems could still readily thrive.

8. Gruchow, journals, February 12, 1987.

9. Gruchow, “Prophet of Joy,” 10.

10. Gruchow, *Necessity of Empty Places*, 124–25.

11. Gruchow, *Grass Roots*, 136.

12. Gruchow, “Kingdom of Grass,” 102.

13. Ibid., 95.

14. Gruchow, *Worlds Within a World*, 19.

15. Gruchow, “Kingdom of Grass,” 76.

16. Sanders, foreword to *Journal of a Prairie Year*, vii–ix.

17. Gruchow, *Journal of a Prairie Year*, 25.

18. For the yeoman landholder, stewardship of natural resources amounted to enlightened self-interest: taking care of watersheds and soil are prudent precautions for ensuring successful harvests in perpetuity.

19. Gruchow, “Kingdom of Grass,” 82.

20. Gruchow, *Grass Roots*, 91.

21. Gruchow, *Boundary Waters*, 57.

22. Gruchow, *Grass Roots*, 86–87.

23. Ibid., 38.

24. Gruchow, *Necessity of Empty Places*, 125.

25. Gruchow, “On the Blue Mound,” 22.

26. Gruchow, *Journal of a Prairie Year*, 71.

27. Gruchow, *Grass Roots*, 46, 15.

28. Gruchow, *Worlds Within a World*, 55.

29. Gruchow, *Boundary Waters*, 38–39.

30. Gruchow, “Naming Names,” 28.

31. Gruchow, *Journal of a Prairie Year*, 71.

32. Ibid., 133.

33. Gruchow, *Grass Roots*, 114.

34. Smith, “American Studies and Environmental History,” 12.

35. Gruchow, “Walking the Border—III,” 18.

36. Gruchow, *Journal of a Prairie Year*, 14.

37. Gruchow, “Meaning of Natural History,” 35.

38. Ibid., 43.

39. Gruchow, *Grass Roots*, 133.

40. Ibid., 21.

41. Gruchow, *Boundary Waters*, 122.
42. Gruchow, *Grass Roots*, 21–22.
43. Gruchow, "On the Blue Mound," 22.
44. Gruchow, "Reasons for Living 9."
45. Gruchow, *Journal of a Prairie Year*, 21.
46. Gruchow, "Kingdom of Grass," 97.
47. Gruchow, *Grass Roots*, 67.
48. Gruchow, *Necessity of Empty Places*, 41.
49. Gruchow, *Grass Roots*, 205.
50. Gruchow, *Necessity of Empty Places*, 53.
51. Gruchow, "On the Blue Mound," 21–22.
52. Gruchow, *Journal of a Prairie Year*, 127.
53. Ibid., 129.
54. Gruchow, *Necessity of Empty Places*, 54–55.
55. Gruchow's projected book about the biological and cultural legacy of bison paralleled the historical scope of Theodore Roethke's planned sequel to "North American Sequence." At the very end of his life, Roethke began to conceive a revisionist account of American history to be composed while visiting the sites of major Native American battles and betrayals, culminating with Wounded Knee.
56. Gruchow, "Kingdom of Grass," 50.
57. Gruchow, *Worlds Within a World*, 36.
58. See *Changes in the Land* by William Cronon and *The Ecological Indian* by Shepard Krech III for an analysis of the uses of fire by Native Americans to reshape woodland and prairie landscapes.
59. Gruchow, *Boundary Waters*, 35.
60. Gruchow, introduction to *The Invisible Pyramid*, vii.
61. Gruchow, *Boundary Waters*, 35.
62. Ibid., 36.
63. Ibid., 30.
64. Gruchow, *Necessity of Empty Places*, 197.
65. Gruchow, journals, June 10, 1994.
66. Gruchow, "Kingdom of Grass," 66.
67. Gruchow, *Worlds Within a World*, 27.
68. Gruchow, *Journal of a Prairie Year*, 46.
69. Ibid., 42.
70. In "Ecology and Memory: The Essays of Paul Gruchow," Louis Martinelli asserts that a single square meter of prairie can sustain up to twenty-five miles of roots.
71. Gruchow, "Kingdom of Grass," 11–12.
72. Gruchow, "What Cranes Say," 17.
73. Gruchow, journals, March 22, 1987.
74. Gruchow, *Grass Roots*, 17.
75. Gustafson, "Revisiting," 7.
76. Martinelli, "Ecology and Memory," 50.
77. Gruchow, *Necessity of Empty Places*, 125.
78. Ibid., 125.
79. Martinelli, "Ecology and Memory," 55.

80. Carlson, "Walking the Two Landscapes," 339.
81. Gruchow, *Boundary Waters*, 201.
82. Gruchow, journals, February 16, 1987.
83. Gruchow, *Boundary Waters*, 189.
84. Gruchow, *Necessity of Empty Places*, 93.
85. Gruchow, introduction to *The Invisible Pyramid*, xii.
86. Gruchow, "Kingdom of Grass," 38.
87. Gruchow, *Uses of Wildness*, 6.
88. Gruchow, journals, 1992.
89. Martinelli, "Ecology and Memory," 63.
90. Gruchow, "Kingdom of Grass," 103.

Epilogue

Returning

Where else is there to go after the land?
To the blue sky? To memory?

—Rick Bass, *The Lost Grizzlies*

The loss of memory will be an eternal loss of meaning.

—Bill McKibben, *The End of Nature*

I first glimpsed for myself a level of natural abundance akin to what had mesmerized early explorers in the Americas—and ultimately lured wave after wave of European settlers—in August of 1982, my first summer of six years living in southeastern Alaska. Solstice had passed just a month before, the longest day of the year. This far north, even after a day's work you could launch a kayak and still get in a good five hours of fishing in full daylight. In fact, at the height of summer, darkness never quite arrives in Juneau. At latitude 58.3019—well *below* the Arctic Circle—sun sets ever so gradually behind the mountains, but before twilight afterglow subsides, sky begins to brighten.

A friend had lent me a thousand dollars to buy a folding kayak perfected by a German firm nearly a century earlier and designed expressly not to roll. I had vowed never to use a canoe that could be swamped all too easily—or worse yet, capsize—in the cold waters of the North Pacific. With its ingenious wooden frame, canvas hull, and built-in inflatable flotation, the Klepper struck me as a far safer choice for sea kayaking. Moreover, "a clever operator," the quaint translation of directions read, could assemble the craft in less time than you might imagine, far faster than I have ever actually managed. In fact, when broken down, the boat could even be checked as luggage on commercial flights. I stowed my kayak all summer at the very end of Thane Road about five miles out of town, tethered to a tree just above the tideline on a neighbor's beach. Max, an elderly gent of Norwegian descent and a lifelong Alaskan, knew these local waters intimately as a commercial fisherman. He suggested that in a day or so I could paddle to a historic Civilian Conservation Corps cabin within sight of the Taku Glacier in the forest near West Turner Lake—stocked decades ago with landlocked sockeye salmon known as kokanee.

After waiting until both weather and tides seemed right, I set out. In a small craft, it is prudent to hug the shoreline; at such latitudes, water temperatures remain cold enough year-round to swiftly trigger hypothermia. To reach the cabin, however, I would need to cross to the far shore of Taku Inlet, traversing open water for a mile or more. It seemed irresistible enough to be worth the risk, and I would not be disappointed. Sheer stone cliffs soared above the southern shore, where in summer colonies of seabirds roost in their untold multitudes, at moments wheeling in impromptu flocks high above. On arrival at the mouth of the river that connects West Turner Lake to the inlet, I found salmon were at the very height of their run, returning like all anadromous fish to spawn in freshwater. The whole spectacle seemed miraculous to me as a newcomer to the north. The channel was so choked with fish that the current turned muscular, raising the entire river, its surface riddled from bank to bank with fins. I realized that this was the same abundance that explorers, colonists, and pioneers had once marveled at across much of the continent. When would I ever get the chance to return and witness this again?

After a soggy night on flats just above tideline at the mouth of the river, the rain let up. Seals feeding just offshore watched me with a curiosity that seemed almost human. I was soon startled to discover that I was not alone. Perched on a rock just upstream was what I least expected, another human being: a semiprofessional wildlife photographer. Like me, he had become entranced watching the inquisitive seals bob and feed. He and two companions from the Upper Midwest had booked the CCC cabin by the lake for a week and arrived by floatplane a few days before. But given dire weather forecasts, they had radioed for a plane to retrieve them that afternoon. They offered me use of the cabin for the following four days, which seemed at the moment a stroke of fate.

While weathered in at West Turner Lake, I holed up in that cabin, keeping a fire burning night and day, reading from the journal that a succession of visitors to this solitary place had added to over the years. They had noticed above all the animals: Sitka black-tailed deer among the huckleberries, seals feasting at the mouth of the river, the inquisitive otter that had watched back, the date each season that salmon had returned, as well as when great flocks of birds had passed through. I had packed provisions lightly, intending to stay just a single night, and was already about out of food. But as my midwestern benefactors explained, custom and etiquette dictated that backcountry pantries be left well stocked in case someone were forced to seek emergency shelter there, such as awaiting evacuation after a bout of hypothermia.

Then again, with berries ripening and salmon spawning, the forest and river were to become my larder. The moment seemed charged with anticipation as I realized that serendipitously I had arrived in the right place at the right time to try my hand for the first time at "living off the land." I soon settled into a routine of tending a fire in the cabin's stone hearth, cooking rice and roe for breakfast, then grilling salmon steaks and filets later in the day, garnished as Alaska Natives had done with oversized huckleberries, delicate thimbleberries, and luminous salmonberries. With this glimpse of subsistence living, I sensed the wisdom of harvesting with restraint, and experienced the primal gratitude that arises from feeding on a diet largely derived from a place by fishing, hunting, and gathering.

In the remote reaches of Taku Inlet in southeastern Alaska that summer decades ago, I glimpsed in the abundance of salmon spawning something ancient, dynamic, and enduring—and understood, as hunting-gathering peoples have always known, that we must ultimately cultivate and sustain a culture of restraint, if the land itself is to last.

At West Turner Lake 8/22/1982

In a cabin on the edge of America's
wilderness, through our lives retreating
like an Ice Age

onto high and barren stone,
I sweep the glaciated granite floor
gathering up the splinters for kindling.

It is an old story here, how conservation
becomes a sacred word beside progress,
how this cabin stands as simple

as the child who wrote in the log,
It's not easy to be a pioneer! Today
I am relearning the ways

of fire, to make it last, how
to build a hearth stone on stone, to lift
a coal from the flames

adding my name to the marks
on these walls, and the year as an
afterthought, when

I was here.

Given the region's notorious tides, I knew return to Juneau by kayak could be treacherous. Timing departure would need to be impeccably coordinated with the tide. Unlike other types of shorelines where tides are generally benign and measured by vertical rise and fall, in fiords, violent currents can arise as massive amounts of water flow through the channel, sometimes creating standing waves and powerful whirlpools. To make matters worse, the route to Juneau would necessitate crossing the width of the inlet again, a mile or more given the contours of each shore. Paddling back from West Turner Lake, I would discover how the forces at work in a tidal fiord can be compounded by erratic weather generated by the surrounding mountains with their sprawling glaciers and towering ice fields. The plan was to glide westward on the outgoing tide until reaching confluence with Gastineau Channel, and turn north just in time to ride rising water the final miles home.

When weather had broken late the day before, after calculating tides once more, I began preparing the boat, and then spent the evening harvesting salmon, fighting the powerful fish to shore until my whole body ached. I clubbed one after another until my heart went numb and sunk them strung together in the cold current until dawn. Smoked and then frozen, this modest catch might last until salmon returned to spawn the following year. Then I tried to sleep. Before light, I rose and began loading the narrow boat, laying salmon down the center of the hull, and, at last, cast off. First light revealed the smooth surface of slack water, that precious lull between incoming and outgoing tides. A thick blanket of fog hung above, just four feet up. In the narrow corridor between, the only muted sounds were water dripping from the paired blades of my paddle, and the wings of a great blue heron cruising at this unaccustomed altitude, its unwavering flight precisely above its own reflection just beneath the cloud.

As I ventured beyond the point and began down the channel, vertical cliffs riddled with nests rose over the near shore, and seabirds again wheeled high above. As outgoing tide had begun to accelerate, I paddled toward the shortest crossing and attempted to angle across the inlet. Current had already begun to surge, tugging forcefully at the rudder even as I edged away from land. By the time I reached the center of the channel, half a mile from either side, it was too far to survive the swim if I were to capsize. A shrill wind began to blow at my back, funneling down the inlet from due east, as waves crested and broke over the bow. The whole boat shuddered—rocked nearly over—then righted itself again. Steering with the surf for the next four hours, I was driven relentlessly seaward, at the utter mercy of wind and waves. Turning crosswise against surging water would have been a fatal mistake.

I believe that three factors saved me that day, every one of them necessary. A spray skirt that I had secured as a precaution when setting out, long before wind had begun to rise, kept breakers from swamping and submerging the kayak. Secondly, the simple flotation technology of the Klepper kept the fragile craft aright among incessant swells. Finally, the weight of salmon in the hold had acted as ballast. I owe my life to their dead weight.

I think that in many ways our planet today faces peril every bit as dire as mine paddling that tiny kayak on the frigid waters of Taku Inlet thirty-five years ago. Environmentally, the stakes have begun to approach the absolute as projected extinctions shock the conscience. Yet it is an old story.
Like John James Audubon, we recognize that survival of wildlife rests on
the preservation of habitat and restraint from heedless harvest.

Like Gene Stratton-Porter, we abhor the serial degradation of self-perpetuating natural places and try to envision their restoration instead.

Like Paul Errington, we seek to protect wetlands that sustain spectacular species biodiversity as well as great migratory flocks.

Like Scott Russell Sanders, we are resolute in hunting for hope and telling the holy.

Like Theodore Roethke, we believe that our spirit is inspired by what might still be thought of as wild.

Like William Stafford, we sense how the legacy of indigenous cultures and environmental integrity are intertwined.

Like Louise Erdrich, we appreciate how Native American languages embody ecological memory as well as sacred, ancient bonds to the places they evolved.

Like Diane Glancy, we grasp that following the loss of ancestral homelands, indigenous peoples must remake their spiritual relationship to place.

Like Elizabeth Dodd, we seek to find clues in the deep past to our future well-being.

Like Paul Gruchow, we recognize that the ethics of agriculture may ultimately dictate the fate of all nature—including our own.

Given the unprecedented ecological crises we presently face, it would be well worth embracing a *single principle* that unites midwestern writers and naturalists across generations—including Audubon, Stratton-Porter, Errington, Leopold, Sanders, and Gruchow—*the sanctity and inviolability of habitat.*

Consider the sort of conservation initiatives supported by their reading of environmental history in the Midwest, remarkably consonant with E. O. Wilson's call for protecting half the planet in as natural a state as still possible:

What if we were to actively conspire with the regenerative forces of *ecological memory*?

What if we were to pledge to protect every *feral edge*, no matter how small a tract, wherever species once indigenous are still vying to return, including buffers bordering cropland?

What if we were to safeguard riparian corridors along every stream and watercourse as well as the remaining ecotones surrounding any wetlands?

What if we were to abate the ecologically devastating consequences of industrial-scale logging practices, particularly clear-cutting, by adopting instead selective timber-harvesting techniques that can enable natural regeneration?

What if efforts at landscape restoration writ large such as major wildlife corridors and reintroduction of buffalo to prairie preserves were complemented by parallel practices on other *open spaces* generally?

What if we were to reimagine every fence line and hedgerow as a *sanctuary*, every roadside and right-of-way as a *refuge*, replacing unnecessary mowing and spraying wherever possible with the reintroduction of native species of plants?

What if local landscape restoration measures that have already met with considerable success were to become the norm at a national level, generating a vast network of *microhabitats* and ecological corridors spanning the continent that could serve as refugia for the future?

Acknowledgments

As with any scholarly project of this scope, each chapter has developed gradually over years and has benefited greatly from the input of others, academic colleagues as well as family and friends, who have served as a cadre of trusted readers throughout the process. I would like to acknowledge here a profound debt to those who have participated in the journey. Their generous comments and suggestions on drafts as well as responses to conference papers have enhanced this work immeasurably. Finally, I wish to credit those journals and presses that have granted permission to reprint material that originally appeared in their publications in an earlier form, as detailed below, and in addition recognize the many professional organizations where related conference papers were presented as this project was being conceived.

First, I would like to thank colleagues in Purdue's College of Liberal Arts who have shown particular interest in and a keen appreciation for this work as it progressed. Nancy Peterson, scholar of Native American literature and former head of the English Department, provided incisive critiques of early versions of the chapter on Louise Erdrich—as did Elizabeth Dodd of Kansas State University, herself subject of chapter 9, "Time's Horizon." Purdue ecocritic Robert Marzec, American Studies scholar Ryan Schneider, and rhetorician Richard Johnson, as well as historians Doug Hurt and Bob May, have all expressed enthusiasm and provided encouragement at various stages of the project. Jill May of the College of Education edited and published my earlier essay on contemporary Native American writers, in many respects the precursor to the discussions of Louise Erdrich and Diane Glancy here. In addition, the Office of the Vice President for Research and Purdue's College of Liberal Arts generously provided funding to underwrite professional indexing of the volume, as well as for me to travel to deliver a keynote address on Paul Gruchow at Texas Tech University, commemorating the recent acquisition of his unpublished papers by the Sowell Family Collection in Literature, Community, and the Natural World.

I am equally grateful to a great many others who generously commented on earlier versions of material. First and foremost, colleagues at the Society for the Study of Midwestern Literature inspired this work with their own, and read drafts of much of this material and responded to related conference papers over the years, especially Bill Barillas, Phil Greasley, Mary Catherine Harper, Marcia Noe, Mary Obuchowski, Ron Primeau, and above all David Radavich. Jon Lauck, cofounder and president of the Midwestern History Association, provided timely guidance and support. Thanks also to site manager emeritus Randy Lehman at the Limberlost Historic Site in Geneva, Indiana, for reviewing an early draft of the chapter on Stratton-Porter's work and to Dave Fox, manager at the Gene Stratton-Porter State Historic Site at Rome City, Indiana, for providing access to the author's private library as well as selected letters.

Kim Stafford has been particularly gracious in responding to several versions of chapter 6, "Landscapes of the Past," which assesses his father's work in relation to ecological and cultural memory. Similarly, Lou Martinelli, founder and director of the Paul Gruchow Foundation, has been a patient and perceptive reader of the manuscript as a whole, as well as a vital guide to my appreciation of Gruchow's signature contributions to the literature of environmental history.

Conference papers presented over the years have been an essential springboard for much of the thinking that underlies this study. The Association for Study of Literature and Environment and the Society for the Study of Midwestern Literature have been particularly valuable venues in this regard, and much of the argumentation in the present volume took shape in their company. Other professional organizations where earlier versions of this work have been delivered and received considered response include the Great Lakes American Studies Association, International Reading Association, Midwest Modern Language Association, and National Council of Teachers of English.

I am also grateful to Lou Martinelli for granting permission to quote materials from the archive of Gruchow's autograph manuscripts now housed in the Sowell Family Collection at Texas Tech University. I am particularly honored to have been the first scholar to be granted access to this substantial body of unpublished work.

A number of my previous publications have been folded into this book. The chapter on Audubon, for example, incorporates, reworks, and expands on material originally published as "The Making of a Conservationist: Audubon's Ecological Memory," which appeared in *Journal of Ecocriticism: Nature, Society, and Literature* in 2013. Similarly, the chapter on Louise Erdrich draws on my article "Landscape and Language in Erdrich's *Books and Islands in Ojibwe Country*," which appeared in *Interdisciplinary Studies in Literature and Environment*, also in 2013.

Thanks to Marcia Noe, chair of the Editorial Committee, for permission to reprint portions of my articles from journals published by the Society for the Study of Midwestern Literature, specifically the following:

"Bearing Witness: Ecological Memory in the Essays of Scott Russell Sanders." *MidAmerica* 39 (2012): 135–44.

"Nature Remains: Scott Russell Sanders's *A Conservationist Manifesto*." *Midwestern Miscellany* 39, no. 1 (Spring 2011): 8–21.

"Writing in History's Wake: Audubon's Portrayal of the Midwest in His Missouri River Journals." *Midwestern Miscellany* 38 (Spring/Fall 2010): 37–53.

"'I'll Be an Indian': Rereading Roethke's 'North American Sequence.' *Midwestern Miscellany* 36 (Spring/Fall 2008): 103–18.

Thanks also to EBSCO for granting permission to reprint text incorporated into the chapter on Theodore Roethke, "Poetics of Place in the North American Sequence," originally published as "Poetics of Place in Roethke's 'North American Sequence,'" in *Critical Insights: American Road Literature*, edited by Ron Primeau, Salem Press, 2013. Used by permission of EBSCO Information Services, Ipswich, MA.

In addition, I wish to acknowledge those at the University of Nevada Press who shepherded this project through the editorial and production phases, above all Director Justin Race for his advocacy since its inception: from recognizing promise in the proposal to providing expert guidance at every juncture. Thanks also to Virginia Fontana for coordinating the overall production process, including the striking cover design, as well as to Delphi photographer Alan McConnell for permission to feature the stunning image of a cataract at Turkey Run in southern Indiana. To view more of his work online, visit the digital gallery at http://www.alanmcconnellphotography.com/. As copy editor, John Mulvihill has been a model of gracious and meticulous professionalism, contributing clarity and impact to the prose. I am also indebted to Michelle Campbell for expertly tackling manuscript preparation tasks, above all the bibliography and a daunting array of notes.

Finally, a number of friends as well as family members have read much of this work in progress and offered input that has led to significant rewriting. Jason Splichal has tracked the trajectory of the manuscript for better than a decade, as have Diane Raven and Michael Scheiwe, and his close readings chapter by chapter have frequently sharpened argumentation. My wife, Julie, has likewise scrutinized drafts year after year, and her thoughtful suggestions have been woven into the finished text. My father has long done the same; indeed, it is at times impossible to distinguish his sensibilities from my own.

Bibliography

Ackerman, Jennifer. *The Genius of Birds*. New York: Penguin, 2016.

Armitage, David. *The Nature Study Movement: The Forgotten Popularizer of America's Conservation Ethic*. Lawrence: University of Kansas Press, 2009.

Audubon, John James. *Audubon and His Journals*. Edited by Maria Audubon. London: J. C. Nimmo, 1898.

———. *Audubon, By Himself: A Profile of John James Audubon*. Edited by Alice Ford. Garden City, NY: American Museum of Natural History, 1969.

———. *The Audubon Reader*. Edited by Richard Rhodes. New York: Knopf, 2006.

———. *Audubon Reader: The Best Writings of John James Audubon*. Edited by Scott Russell Sanders. Bloomington: Indiana University Press, 1986.

———. *Audubon's Birds of America: The Audubon Society Baby Elephant Folio*. New York: Abbeville Press, 1991.

———. *The Birds of America*. New York: Macmillan, 1941.

———. *The 1826 Journal of John James Audubon*. Edited by Alice Ford. New York: Abbeville Press, 1987.

———. *John James Audubon: Writings and Drawings*. Edited by Christoph Irmsher. New York: Library of America, 1999.

———. *The Missouri River Journals of John James Audubon*. Edited by Daniel Patterson. Lincoln: University of Nebraska Press, 2016.

———. *Ornithological Biography*. Edinburgh: Adam Black, 1831.

———. *The Viviparous Quadrupeds of North America*. New York: Wellfleet, 2005.

Backes, David. *A Wilderness Within: The Life of Sigurd F. Olson*. Minneapolis: University of Minnesota Press, 1999.

Bakeless, John. *America as Seen by Its First Explorers*. Mineola, NY: Dover, 1950.

Balakian, Peter. *Theodore Roethke's Far Fields: The Evolution of His Poetry*. Baton Rouge: Louisiana State University Press, 1989.

Barillas, William. *The Midwestern Pastoral: Place and Landscape in Literature of the American Heartland*. Athens: Ohio University Press, 2006.

Bass, Rick. *The Lost Grizzlies: A Search for Survivors in the Wilderness of Colorado*. New York: Houghton Mifflin, 1995.

Bearss, Linda, and Ronald Primeau, eds. *Midwestern Miscellany* 36 (Being a Collection of Essays on Theodore Roethke) (2008).

Begossi, Alpina. "Use of Ecological Methods in Ethnobotany: Diversity Indices." *Economic Botany* 50, no. 3 (1996): 280–89.

Bent, Arthur Cleveland. *Life Histories of North American Jays, Crows, and Titmice*. Mineola, NY: Dover, 1988.

Bergthaller, Hannes. "Introduction: Ecocriticism and Environmental History." *Interdisciplinary Studies in Literature and Environment* 22, no. 1 (2015): 5–8.

Berkes, Fikret. *Sacred Ecology*. New York: Routledge, 1999.

Berry, Wendell. "Ignorance." In *Science and the Myth of Progress*, edited by Merdad M. Zarandi and Giovanni Monastra. Lanham, MD: World Wisdom, 2003.

———. *The Unsettling of America: Culture and Agriculture*. San Francisco: Sierra Club Books, 1977.

BirdLife International 2012. "*Pinguinus impennis.*" *The IUCN Red List of Threatened Species*. Version 2012.2. Accessed June 11, 2013. www.iucnredlist.org.

Birkelo, Cheryl. "Gene Stratton-Porter: Scholar of the Natural World in *A Girl of the Limberlost*." *Midwestern Miscellany* 49 (Spring 2012): 7–29.

Birkhead, Tim. *The Wisdom of Birds: An Illustrated History of Ornithology*. New York: Bloomsbury, 2008.

Boehme, Sarah, ed. *John James Audubon in the West: The Last Expedition and Mammals of North America*. New York: Abrams, 2000.

Bowers, Neal. *Theodore Roethke: The Journey from I to Otherwise*. Columbia: University of Missouri Press, 1982.

Branch, Michael. "Indexing American Possibilities: The Natural History Writing of Bartram, Wilson, and Audubon." In *The Ecocriticism Reader: Landmarks in Literary Ecology*, edited by Cheryll Glotfelty and Harold Fromm, 282–302. Athens: University of Georgia Press, 1996.

Branch, Michael, and Daniel Philippon. *The Height of the Mountains: Writing from Virginia's Blue Ridge Mountains and Shenandoah Valley*. Baltimore: Johns Hopkins University Press, 1998.

Buell, Lawrence. *The Environmental Imagination: Thoreau, Nature Writing, and the Formation of American Culture*. Cambridge, MA: Belknap Press of Harvard University Press, 1996.

Bunge, Nancy. "A Mystical Midwest: Louise Erdrich's *Tracks* and the Midwestern Tradition." *MidAmerica* 35 (2008): 106–14.

Callicott, J. Baird, ed. *Companion to* A Sand County Almanac: *Interpretive and Critical Essays*. Madison: University of Wisconsin Press, 1987.

———. *In Defense of the Land Ethic: Essays in Environmental Philosophy*. Albany: State University of New York Press, 1989.

Carlson, Douglas. "Walking the Two Landscapes: An Essay Review." *Georgia Review* 52, no. 2 (1998): 334–40.

Catlin, George. *Letters and Notes on the Manners, Customs and Conditions of North American Indians*. Mineola, NY: Dover, 1973.

———. *North American Indians*. Edited by Peter Matthiessen. New York: Penguin, 1989.

Chansiguad, Valerie. *The History of Ornithology*. London: New Holland, 2009.

Cohen, Michael P. "Blues in the Green: Ecocriticism Under Critique." *Environmental History* 9, no. 1 (2004): 9–36.

Cokinos, Christopher. *Hope Is the Thing with Feathers: A Personal Chronicle of Vanished Birds*. New York: J. P. Tarcher, 2000.

Colter, Susan. "Born Free: America's Largest Dam Removal Begins in the Pacific Northwest." *Audubon* (January–February 2012): 17.

Comstock, Anna Botsford. *Handbook of Nature Study*. Ithaca, NY: Cornell University Press, 1911.

Cronon, William. *Changes in the Land: Indians, Colonists, and the Ecology of New England*. 2nd ed. New York: Hill and Wang, 2003.

Dean, Thomas, ed. *The Grace of Grass and Water: Writing in Honor of Paul Gruchow*. North Liberty, IA: Ice Cube Press, 2007.

DeVoto, Bernard. *Across the Wide Missouri*. Boston: Houghton Mifflin, 1947.

DiBenedetto, David. *On the Run: An Angler's Journey Down the Striper Coast*. New York: William Morrow, 2003.

Dodd, Elizabeth. *Archetypal Light*. Reno: University of Nevada Press, 2001.

———. *Horizon's Lens: My Time on the Turning World*. Lincoln: University of Nebraska Press, 2012.

———. *In the Mind's Eye: Essays across the Animate World*. Lincoln: University of Nebraska Press, 2008.

———. *Like Memory, Caverns*. New York: New York University Press, 1992.

———. *Prospect: Journeys and Landscapes*. Lincoln: University of Nebraska Press, 2003.

Dougherty, James. "Theodore Roethke's 'North American Sequence': Religious Awakening in the West." *Literature and Belief* 21 (2001): 177–203.

Dreese, Donelle. *Ecocriticism: Creating Self and Place in Environmental and American Indian Literature*. New York: Peter Lang, 2002.

Durant, Mary, and Michael Harwood. *On the Road with John James Audubon*. New York: Dodd, Mead, 1980.

Ehle, John. *Trail of Tears: The Rise and Fall of the Cherokee Nation*. New York: Random House, 1988.

Eiseley, Loren. *The Immense Journey*. New York: Random House, 1959.

———. *The Night Country*. New York: Scribner, 1971.

———. *The Star Thrower*. New York: Times Books, 1978.

Eisler, Benita. *The Red Man's Bones: George Catlin, Artist and Showman*. New York: Norton, 2013.

Erdrich, Louise. "Big Grass." In *Heart of the Land: Essays on Last Great Places*, edited by Joseph Barbato and Lisa Weinerman, 145–50. New York: Pantheon, 1994.

———. *Books and Islands in Ojibwe Country*. Washington, DC: National Geographic, 2003.

———. *Original Fire: Selected and New Poems*. New York: Harper, 2003.

Errington, Paul. *Muskrats and Marsh Management*. Lincoln: University of Nebraska Press, 1961.

———. *Of Men and Marshes*. New York: Macmillan, 1957.

———. *Of Predation and Life*. Iowa City: University of Iowa Press, 1967.

———. *Of Wilderness and Wolves*. Iowa City: University of Iowa Press, 2015.

———. *A Question of Values*. Iowa City: University of Iowa Press, 1987.

———. *The Red Gods Call*. Iowa City: University of Iowa Press, 1973.

Ewers, John. *George Catlin: Painter of Indians of the West*. Washington, DC: Smithsonian, 1955.

Felstiner, John. *Can Poetry Save the Earth? A Field Guide to Nature Poems*. New Haven, CT: Yale University Press, 2009.

Finch, Robert. "Introduction: The Delights and Dilemmas of *A Sand County Almanac*." In *A Sand County Almanac*, by Aldo Leopold, xv–xxvii. New York: Oxford University Press, 1987.

Flake, Lester, and Alan Wentz. "'Of Men and Marshes' Symposium and Dedication of Paul L. Errington Memorial Marsh." *Wildlife Society Bulletin* 4, no. 4 (1976): 183–85.

Flannery, Timothy. *The Eternal Frontier: An Ecological History of North America and Its Peoples*. New York: Atlantic Monthly Press, 2001.

Ford, Alice. *John James Audubon: A Biography*. New York: Abbeville Press, 1988.

Foster, Lance M. "The Ioway and the Landscape of Southeast Iowa." *Journal of the Iowa Archaeological Society* 43 (1996): 1–5. http://ioway.nativeweb.org/iowaylibrary/seiowa.htm.

———. "*Tanji na Che:* Recovering the Landscape of the Ioway." In *Recovering the Prairie*, edited by Robert F. Sayre. Madison: University of Wisconsin Press, 1999. http://ioway.nativeweb.org/iowaylibrary/tanji.htm.

Francaviglia, Richard. *Believing in Place: A Spiritual Geography of the Great Basin*. Reno: University of Nevada Press, 2003.

Franklin, Jon. *The Wolf in the Parlor: The Eternal Connection Between Humans and Dogs*. New York: Henry Holt and Co., 2009.

Fritzell, Peter. "The Conflicts of Ethical Conscience." In *Companion to* A Sand County Almanac: *Interpretive and Critical Essays*, edited by J. Baird Callicott, 128–53. Madison: University of Wisconsin Press, 1987.

Furtman, Michael. *Magic on the Rocks: Canoe Country Pictographs*. Duluth, MN: Birch Portage Press, 2000.

Glancy, Diane. *Pushing the Bear: After the Trail of Tears*. Norman: University of Oklahoma Press, 2009.

———. *Pushing the Bear: A Novel of the Trail of Tears*. New York: Harcourt, 1996.

———. *Stone Heart: A Novel of Sacajawea*. Woodstock, NY: Overlook Press, 2003.

Glotfelty, Cheryll, and Harold Fromm, eds. *The Ecocriticism Reader: Landmarks in Literary Ecology*. Athens: University of Georgia Press, 1996.

Green, Amy S. "Two Women Naturalists and the Search for Autonomy: Anna Botsford Comstock and the Producer Ethic; Gene Stratton-Porter and the

Gospel of Wealth." *Women's Studies Quarterly* 29, no. 1/2 (Summer 2001): 145–54.

Gruchow, Paul. *Boundary Waters: The Grace of the Wild*. Minneapolis: Milkweed Editions, 1997.

———. *Grass Roots: The Universe of Home*. Minneapolis: Milkweed Editions, 1995.

———. Introduction to *The Invisible Pyramid*, by Loren Eiseley. Lincoln: University of Nebraska Press, 1998.

———. *Journal of a Prairie Year*. 1985. Reprint, Minneapolis: Milkweed Editions, 2009.

———. Journals. Paul Gruchow Papers.

———. "The Kingdom of Grass." Paul Gruchow Papers.

———. *Letters to a Young Madman: A Memoir*. Minneapolis: Levins, 2012.

———. "The Meaning of Natural History." In Dean, *The Grace of Grass and Water*, 21–46.

———. "Naming Names: A New List of Minnesota's Endangered Species Call All of Us to Account." *Minnesota Monthly*, March 1989, 23–28.

———. *The Necessity of Empty Places*. New York: St. Martin's Press, 1988.

———. "On the Blue Mound: Visiting the Ghosts of the Lost World of the Prairies." *Minnesota Monthly*, July 1986, 20–23.

———. Paul Gruchow Papers. Southwest Collection/Special Collections Library. Texas Tech University, Lubbock, TX.

———. "The Prophet of Joy," *Sigurd F. Olson 1899–1982*. Minneapolis: Friends of the Boundary Waters Wilderness Foundation and The Wilderness Society, 1999.

———. "Reasons for Living 9." Paul Gruchow Papers.

———. *Travels in Canoe Country*. Boston: Bulfinch Press, 1992.

———. *The Uses of Wildness*. Minneapolis: Boundary Waters Wilderness Foundation, 1988.

———. "Walking the Border—III." *Minnesota Monthly*, February 1989, 18–21.

———. "What Cranes Say." *Minnesota Monthly*, March 1987, 12–17.

———. *Worlds Within a World: Reflections on Visits to Minnesota Scientific and Natural Area Preserves*. Saint Paul: Minnesota Department of Natural Resources, 1999.

Gruchow, Paul, et al. *Prairie Roots: 2001 Harvest Symposium*. North Liberty, IA: Ice Cube Press, 2001.

Gustafson, Loren. "Revisiting Paul Gruchow's Back Roads." *Whitewater River Letter* 1, no. 2 (2009): 5–11.

Habegger, Larry, and Amy Carlson, eds. *The Gift of Birds: True Encounters with Avian Spirits*. San Francisco: Travelers' Tales, 1999.

Halifax, Joan. *The Fruitful Darkness*. New York: Grove Press, 2004.

Halpin, Marjorie. Introduction to *Letters and Notes on the Manners, Customs, and Conditions of North American Indians*. Mineola, NY: Dover, 1973.

Harding, Walter, and Michael Meyer. *The New Thoreau Handbook*. New York: New York University Press, 1980.

Hart-Davis, Duff. *Audubon's Elephant: America's Greatest Naturalist and the Making of the Birds of America*. New York: Henry Holt and Co., 2004.

Hassrick, Royal B. *The George Catlin Book of American Indians*. New York: BBS, 1994.

Herrick, Francis. *Audubon the Naturalist: A History of His Life and Time*. 1917. Reprint, Mineola, NY: Dover, 1968.

Hicks, Brian. *Toward the Setting Sun: John Ross, the Cherokees, and the Trail of Tears*. New York: Atlantic Monthly Press, 2011.

Hill, Beth, and Ray Hill. *Indian Petroglyphs of the Pacific Northwest*. Seattle: University of Washington Press, 1974.

Homan, Tim, ed. *A Yearning for Wildness: Environmental Quotations from the Writings of Henry David Thoreau*. Atlanta: Peachtree, 1991.

Hudson, John. *Across This Land: A Regional Geography of the United States and Canada*. Baltimore: Johns Hopkins University Press, 2002.

———. *Crossing the Heartland: Chicago to Denver*. New Brunswick, NJ: Rutgers University Press, 1992.

———. *Making the Corn Belt: A Geographical History of Middle-Western Agriculture*. Bloomington: Indiana University Press, 1994.

Huhndorf, Shari. "'Mapping by Words': The Politics of Land in Native American Literature." In *The Cambridge Companion to Literature and the Environment*, edited by Louise Westling, 45–58. Cambridge: Cambridge University Press, 2014.

Hurd, Barbara. *Stirring the Mud: On Swamps, Bogs, and Human Imagination*. Boston: Beacon Press, 2001.

Indiana Department of Natural Resources. *Prophetstown State Park Interpretive Plan*. Indianapolis: Indiana Department of Natural Resources, 2004.

Jackson, Marion. *The Natural Heritage of Indiana*. Bloomington: Indiana University Press, 1997.

Jacobs, Alan. *Thoreau: Transcendent Nature for a Modern World*. London: Watkins, 2012.

John James Audubon Museum, Henderson, Kentucky. http://parks.ky.gov/find parks/recparks/au/.

Kalaidjian, Walter. *Understanding Theodore Roethke*. Columbia: University of South Carolina Press, 1987.

Kearful, Frank. "Regions of the Self: Theodore Roethke's 'North American Sequence.'" *Modern American Poetry*, March 20, 2012. http://www.english .illinois.edu/maps/ poets/m_r/roethke/kearful.htm.

Kilcup, Karen. *Fallen Forests: Emotion, Embodiment, and Ethics in American Women's Environmental Writing, 1781–1924*. Athens: University of Georgia Press, 2013.

King, Rollin Patterson. *Gene Stratton-Porter: A Lovely Light*. Chicago: Adams Press, 1979.

Knoeller, Christian. "Bearing Witness: Ecological Memory in the Essays of Scott Russell Sanders." *MidAmerica* 39 (2012): 135–44.

———. "'I'll Be an Indian': Rereading Roethke's 'North American Sequence.'" *Midwestern Miscellany* 36 (Spring/Fall 2008): 103–18.

———. "Landscape and Language in Erdrich's *Books and Islands in Ojibwe Country*." *ISLE* 19, no. 4 (2013): 645–60.

———. "The Making of a Conservationist: Audubon's Ecological Memory." *Journal of Ecocriticism: Nature, Society and Literature* 5, no. 2 (July 2013): 1–18.

———. "Nature Remains: Scott Russell Sanders's *A Conservationist Manifesto*." *Midwestern Miscellany* 39 (Spring 2011): 8–21.

———. "Poetics of Place in Theodore Roethke's 'North American Sequence.'" *American Road Literature*, edited by Ronald Primeau, 190–211. Ipswich, MA: Salem Press, 2013.

———. "Preface: A Tribute to Scott Russell Sanders." *Midwestern Miscellany* 39 (Fall 2011): 4–5.

———. "Writing in History's Wake: Audubon's Portrayal of the Midwest in His *Missouri River Journal*." *Midwestern Miscellany* 38 (Spring/Fall 2010): 37–53.

Knott, John. *Imagining Wild America*. Ann Arbor: University of Michigan Press, 2002.

Kohler, Robert. *All Creatures: Naturalists, Collectors, and Biodiversity, 1850–1950*. Princeton, NJ: Princeton University Press, 2006.

———. "Paul Errington, Aldo Leopold, and Wildlife Ecology: Residential Science." *Historical Studies in the Natural Sciences* 41, no. 2 (2011): 216–54.

Krech, Shepard. *The Ecological Indian: Myth and History*. New York: Norton, 2000.

La Belle, Jenijoy. *The Echoing Wood of Theodore Roethke*. Princeton, NJ: Princeton University Press, 1976.

Lao Tzu. *The Way of Life According to Lao Tzu*. Translated by Witter Bynner. New York: Capricorn, 1962.

Leach, William. *Butterfly People: An American Encounter with the Beauty of the World*. New York: Pantheon, 2013.

Lentz, David L. *Imperfect Balance: Landscape Transformations in the Precolumbian Americas*. New York: Columbia University Press, 2000.

Leopold, Aldo. *Game Management*. New York: Scribner, 1933.

———. *A Sand County Almanac: And Sketches Here and There*. New York: Oxford University Press, 1949.

Lincoln, Kenneth. *Sing with the Heart of a Bear: Fusions of Native and American Poetry, 1890–1999*. Berkeley: University of California Press, 2000.

Lindsey, Alton. *The Bicentennial of John James Audubon*. Bloomington: Indiana University Press, 1985.

Lischke, Ute, and David McNab, eds. *The Long Journey of a Forgotten People: Métis Identities and Family Histories*. Waterloo, ON: Wilfrid Laurier University Press, 2007.

———. "Reflections on Metis Connections in the Life and Writings of Louise Erdrich." In Lischke and McNab, *The Long Journey of a Forgotten People*, 39–53.

Long, Judith. *Gene Stratton-Porter: Novelist and Naturalist*. Indianapolis: Indiana Historical Society, 1990.

Lopez, Barry. *About This Life: Journeys on the Threshold of Memory*. New York: Vintage, 1998.

———, ed. *Home Ground: Language for an American Landscape*. New York: Trinity University Press, 2006.

———. *Of Wolves and Men*. New York: Scribner, 1978.

Low, Denise. Introduction to *The Kansas Poems of William Stafford*. Topeka, KS: Woodley Memorial Press, 1990.

Lynch, Tom. "Ecopastoralism: Settler Colonial Pastoral Imaginary in the US West and Australian Outback." *Australian Literary Studies* 30, no. 2 (2015): 145–58.

Lynch, Tom, and Susan N. Maher, eds. *Artifacts and Illuminations: Critical Essays on Loren Eiseley*. Lincoln: University of Nebraska Press, 2012.

———. "The Borders Between Us: Loren Eiseley's Ecopoetics." In Lynch and Maher, *Artifacts and Illuminations*, 123–42.

Maher, Susan N. *Deep Map Country: Literary Cartography of the Great Plains*. Lincoln: University of Nebraska Press, 2014.

———. "The Places Below: Mapping the Invisible Universe in Loren Eiseley's Plains Essays." In Lynch and Maher, *Artifacts and Illuminations*, 55–76.

Malkoff, Karl. *Theodore Roethke: An Introduction to the Poetry*. New York: Columbia University Press, 1966.

Mancall, Peter. *Envisioning America: English Plans for the Colonization of North America, 1580–1640*. Boston: Bedford/St. Martin's, 1995.

Manley Hopkins, Gerard. *Selected Poems of Gerard Manley Hopkins*. Edited by Bob Blaisdell. Mineola, NY: Dover, 2011.

Manning, Richard. *Grassland: The History, Biology, Politics, and Promise of the American Prairie*. New York: Penguin, 1995.

Marinelli, Janet. "Native, or Not So Much?" *National Wildlife*, June–July 2016, 12–14.

Marsh, George Perkins. *Man and Nature: Or, Physical Geography as Modified by Human Action*. 1865. Seattle: University of Washington Press, 2003.

Martinelli, Louis. "Ecology and Memory: The Essays of Paul Gruchow." In Dean, *The Grace of Grass and Water: Writing in Honor of Paul Gruchow*, 47–65.

Marx, Leo. *The Machine in the Garden: Technology and the Pastoral Ideal in America*. New York: Oxford University Press, 1964.

Matthiessen, Peter. "The Last Cranes of Siberia." In *The Gift of Birds: True Encounters with Avian Spirits*, edited by Larry Habegger and Amy Carlson, 275–86. San Francisco: Travelers' Tales, 1999.

———. *The Snow Leopard*. New York: Viking, 1978.

———. *Wildlife in America*. New York: Viking, 1959.
McClintock, James. *Nature's Kindred Spirits*. Madison: University of Wisconsin Press, 1994.
McCracken, Harold. *George Catlin and the Old Frontier: A Biography and Picture Gallery of the Dean of Indian Painters*. 1959. Reprint, New York: Bonanza, 1984.
McDermott, John Francis, ed. *Audubon in the West*. Norman: University of Oklahoma Press, 1965.
———, ed. *Up the Missouri with Audubon: The Journal of Edward Harris*. Norman: University of Oklahoma Press, 1951.
McHugh, Tom. *The Time of the Buffalo*. New York: Knopf, 1972.
McKibben, Bill. *The End of Nature*. New York: Random House, 1989.
———. "From *The End of Nature*." In *Nature Writing: The Tradition in English*, edited by Robert Finch and John Elder, 1120–30. New York: Norton, 2002.
McNab, David. "The Significance of Place and Indigenous Knowledge in Louise Erdrich's *Books and Islands in Ojibwe Country*." *European Journal of American Culture* 25, no. 3 (2006): 205–19.
Meehan, Jeannette Porter. *The Lady of the Limberlost: The Life and Letters of Gene Stratton-Porter*. Garden City, NY: Doubleday, Doran, 1928.
Mercier, Stephen. "Epic Narratives of Evolution: John Burroughs and Loren Eiseley." In Lynch and Maher, *Artifacts and Illuminations*, 251–70.
Merwin, W. S. *The Compass Flower: Poems*. New York: Antheneum, 1977.
———. "W. S. Merwin, The Art of Poetry, No. 38." Interview by Edward Hirsch. *Paris Review*, no. 102 (Spring 1987). https://www.theparisreview.org/interviews/2692/w-s-merwin-the-art-of-poetry-no-38-w-s-merwin.
Mills, Ralph. "In the Way of Becoming: Roethke's Last Poems." In *Theodore Roethke: Essays on the Poetry*, edited by Arnold Stein, 115–35. Seattle: University of Washington Press, 1965.
Momaday, N. Scott. *The Way to Rainy Mountain*. Albuquerque: University of New Mexico Press, 1969.
Moore, Kathleen Dean. *Holdfast: At Home in the Natural World*. Guilford, CT: Lyons Press, 1999.
Nazarea, Virginia. "Local Knowledge and Memory in Biodiversity Conservation." *Annual Review of Anthropology* 35 (2006): 317–35.
Nelson, Cary. "An Essay by Cary Nelson on 'North American Sequence,' from *Our Last First Poets: Vision and History in Contemporary American Poetry*. Champaign: University of Illinois Press, 1981. http://www.english.illinois.edu/maps/poets/m_r/roethke/nelson.htm.
Nelson, Richard. *Make Prayers to the Raven: A Koyukon View of the Northern Forest*. Chicago: University of Chicago Press, 1986.
Nelson, Robert. *Place and Vision: The Function of Landscape in Native American Fiction*. New York: Peter Lang, 1993. https://facultystaff.richmond.edu/~rnelson/PandV/front.html.

Neruda, Pablo. *Full Woman, Fleshy Apple, Hot Moon*. Translated by Stephen Mitchell. New York: HarperCollins, 1997.

Nicholls, Steve. *Paradise Found: Nature in America at the Time of Discovery*. Chicago: University of Chicago Press, 2009.

Nordstrom, Lars. *Theodore Roethke, William Stafford, and Gary Snyder: The Ecological Metaphor as Transformed Regionalism*. PhD diss., Uppsala: Upsaliensis S. Academiae, 1989.

Novak, Barbara. *Nature and Culture: American Landscape and Painting 1825–1875*. New York: Oxford University Press, 1980.

Nye, Naomi Shihab. Introduction to *The Way It Is: New and Selected Poems*, by William Stafford. Saint Paul, MN: Graywolf Press, 1999.

Obuchowski, Mary DeJong. *Field O' My Dreams: The Poetry of Gene Stratton-Porter*. Kent, OH: Kent State University Press, 2007.

Olson, Sigurd. *Reflections from the North Country*. 1976. Reprint, Minneapolis: University of Minnesota Press, 1998.

Osborne, Terry. Review of *The Rediscovery of the Wild*, edited by Peter H. Kahn Jr. and Patricia H. Hasbach. *Interdisciplinary Studies in Literature and Environment* 20, no. 1 (2014): 210–11.

Paddock, Joe. *Keeper of the Wild: The Life of Ernest Oberholtzer*. Saint Paul: Minnesota Historical Society, 2001.

Parini, Jay. *Theodore Roethke: An American Romantic*. Amherst: University of Massachusetts Press, 1979.

Patterson, Daniel, ed. *The Missouri River Journals of John James Audubon*. Lincoln: University of Nebraska Press, 2016.

Pauketat, Timothy. *Cahokia: Ancient America's Great City on the Mississippi*. New York: Penguin, 2010.

Paxton, Robert. "Birds: The Inner Life." *New York Review of Books*, January 10, 2013, 30–32.

Paz, Octavio. *The Collected Poems of Octavio Paz, 1957–1987*. Translated by Eliot Weinberger. New York: New Directions, 1991.

Peattie, Donald. *Audubon's America: The Narratives and Experiences of John James Audubon*. Boston: Houghton, 1940.

Peck, Robert. "Audubon and Bachman: A Collaboration in Science." In Boehme, *John James Audubon in the West*, 71–118.

Perdue, Theda, and Michael D. Green. *The Cherokee Removal: A Brief History with Documents*. 3rd ed. Boston: Bedford/St. Martins, 2016.

Peterson, Nancy. *People of the Troubled Water: A Missouri River Journal*. Frederick, CO: Renaissance, 1988.

Philippon, Daniel J. *Conserving Words: How American Nature Writers Shaped the Environmental Movement*. Athens: University of Georgia Press, 2004.

Plum, Sydney Landon. *Coming Through the Swamp: The Nature Writing of Gene Stratton Porter*. Salt Lake City: University of Utah Press, 1996.

Price, John. *Not Just Any Land: A Personal and Literary Journey into the American Grasslands*. Lincoln: University of Nebraska Press, 2004.

Pritchard, James, et al. "The Landscape of Paul Errington's Work." *Wildlife Society Bulletin* 34, no. 5 (2006): 1411–16.

Proctor, Jonathon. "Defender to the Core." *Defenders* 86.4, no. 13. Washington, DC: Defenders of Wildlife, 2011.

Pruvost, Melanie, et al. "Genotypes of Predomestic Horses Match Phenotypes Painted in Paleolithic Works of Cave Art." *Proceedings of the National Academy of Sciences* 108, no. 46 (2011): 626–30.

Quechenbach, Bernard. *Beyond the Far Field: American Nature Poetry in the Late Twentieth Century*. Charlottesville: University Press of Virginia, 2000.

Rhodes, Richard. *John James Audubon: The Making of an American*. New York: Knopf, 2004.

Richards, Bertrand. *Gene Stratton Porter*. Boston: Twayne, 1980.

Roethke, Theodore. "An American Poet Introduces Himself and His Poems." In Roethke, *On the Poet and His Craft*, 7–13.

———. *Collected Poems*. New York: Doubleday, 1966.

———. *The Far Field*. New York: Doubleday, 1964.

———. *On the Poet and His Craft: Selected Prose of Theodore Roethke*. Edited by Ralph J. Mills Jr. Seattle: University of Washington Press, 1968.

———. "On 'Identity.'" In Roethke, *On the Poet and his Craft*, 18–27.

———. *The Selected Letters of Theodore Roethke*. Edited by Ralph J. Mills Jr. Seattle: University of Washington Press, 1968.

———. *Straw for the Fire: From the Notebooks of Theodore Roethke, 1943–1963*. Edited by David Wagoner. New York: Doubleday, 1972.

———. "Theodore Roethke." In Roethke, *On the Poet and his Craft*, 14–17.

Ronda, James. *Lewis and Clark Among the Indians*. Lincoln: University of Nebraska Press, 2002.

Rosenthal, M. L. *The New Poets: American and British Poetry Since World War II*. London: Oxford University Press, 1965.

Rossi, William, ed. Introduction to *"Wild Apples" and Other Natural History Essays*, by Henry David Thoreau. Athens: University of Georgia Press, 2002.

Russell, Howard. *Indian New England Before the Mayflower*. Hanover, NH: University Press of New England, 1980.

Sanders, Scott Russell. *A Conservationist Manifesto*. Bloomington: Indiana University Press, 2009.

———. *The Country of Language*. Minneapolis: Milkweed Editions, 1999.

———. *Earth Works: Selected Essays*. Bloomington: Indiana University Press, 2012.

———. Foreword to *Journal of a Prairie Year*, by Paul Gruchow, vii–ix. Minneapolis: Milkweed Editions, 2009.

———. *Hunting for Hope: A Father's Journey*. Boston: Beacon, 1998.

———. *A Private History of Awe*. New York: North Point Press, 2006.
———. *Staying Put: Making a Home in a Restless World*. Boston: Beacon, 1993.
———. *Wilderness Plots*. New York: Morrow, 1983.
———. *Writing from the Center*. Bloomington: Indiana University Press, 1995.
Sayre, Robert F., ed. *Recovering the Prairie*. Madison: University of Wisconsin Press, 1999.
Schaller, George. "Politics Is Killing the Big Cats." *National Geographic*, December 2011, 89–91.
Schorger, A. W. "In Memoriam: Paul Lester Errington." *The Auk* 83, no. 1 (1966): 52–65.
Schweninger, Lee. *Listening to the Land: Native American Literary Responses to the Landscape*. Athens: University of Georgia Press, 2008.
Seager, Allan. *The Glass House: The Life of Theodore Roethke*. New York: McGraw-Hill, 1968.
Shepherd, Paul. *The Tender Carnivore and the Sacred Game*. Athens: University of Georgia Press, 1998.
Silver, Timothy. *Mount Mitchell and the Black Mountains: An Environmental History of the Highest Peaks in Eastern America*. Chapel Hill: University of North Carolina Press, 2003.
———. *A New Face on the Countryside: Indians, Colonists, and Slaves in South Atlantic Forests, 1500–1800*. Cambridge: Cambridge University Press, 1990.
Sivils, Matthew. "Paul L. Errington: His Life and Work." Introduction to *Of Men and Marshes*, by Paul Errington. Iowa City: University of Iowa Press, 2012.
Slovic, Scott. "Editor's Note." *Interdisciplinary Studies in Literature and Environment* 22, no. 1 (2015): 1–3.
———. "Nature Writing and Environmental Psychology: The Interiority of Outdoor Experience." In *The Ecocriticism Reader: Landmarks in Literary Ecology*, edited by Cheryll Glotfelty and Harold Fromm, 351–70. Athens: University of Georgia Press, 1996.
———. "Scott Russell Sanders: A Portrait." In *The Country of Language*, by Scott Russell Sanders, 89–106. Minneapolis: Milkweed Editions, 1999.
———. *Seeking Awareness in American Nature Writing*. Salt Lake City: University of Utah Press, 1998.
Smith, Chadwick, and Rennard Stickland. *Building One Fire: Art and World View in Cherokee Life*. Tahlequah, OK: Cherokee Nation, 2010.
Smith, Kendra. "American Studies and Environmental History: Provoking Thought and Practice." *American Studies* 46, no. 1 (2005): 5–22.
Snodgrass, W. D. "'The Anguish of Concreteness': Theodore Roethke's Career." In *Theodore Roethke: Essays on the Poetry*, edited by Arnold Stein, 78–93. Seattle: University of Washington Press, 1965.
———. "The Last Poems of Theodore Roethke." *New York Review of Books*, October 8, 1964, 5–6.

Souder, William. *Under a Wild Sky: John James Audubon and the Making of Birds of America*. New York: North Point Press, 2004.

Stafford, Kim. *Early Morning: Remembering My Father, William Stafford*. Saint Paul, MN: Graywolf Press, 2003.

Stafford, William. *Allegiances*. New York: Harper & Row, 1970.

———. *Another World Instead: The Early Poems of William Stafford, 1937–1947*. Edited by Frank Marchant. Saint Paul, MN: Graywolf Press, 2008.

———. *The Darkness Around Us Is Deep*. Edited by Robert Bly. New York: Harper Perennial, 1993.

———. *A Glass Face in the Rain*. New York: Harper & Row, 1982.

———. *The Kansas Poems of William Stafford*. Topeka, KS: Woodley Memorial Press, 1990.

———. *My Name Is William Tell*. Lewiston, ID: Confluence Press, 1992.

———. *An Oregon Message*. New York: Harper & Row, 1987.

———. *Passwords*. New York: Harper Perennial, 1991.

———. *The Rescued Year*. New York: Harper & Row, 1966.

———. *Smoke's Way*. Port Townsend, WA: Graywolf Press, 1983.

———. *Someday, Maybe*. New York: Harper & Row, 1973.

———. *Sound of the Ax: Aphorisms and Poems by William Stafford*. Edited by Frank Marchant. Pittsburgh, PA: University of Pittsburgh Press, 2014.

———. *Stories That Could Be True*. New York: Harper & Row, 1977.

———. *Traveling Through the Dark*. New York: Harper & Row, 1962.

———. *The Way It Is: New and Selected Poems*. Saint Paul, MN: Graywolf Press, 1999.

———. *West of Your City*. Los Gatos, CA: Talisman Press, 1960.

———. "Willingly Local: A Conversation with William Stafford About Regionalism and Northwest Poetry." By Lars Nordstrom. Lake Oswego, Oregon, December 20, 1985. http://www.worldpapercat.com/1276/Article 1466575.htm.

Staples, Hugh. "Rose in the Sea-Wind: A Reading of Theodore Roethke's 'North American Sequence.'" *American Literature* 6 (May 1964): 189–203.

Stein, Arnold, ed. *Theodore Roethke: Essays on the Poetry*. Seattle: University of Washington Press, 1965.

Stirrup, David. *Louise Erdrich*. Manchester: Manchester University Press, 2010.

Stolzenburg, William. "The Long Rangers." *Nature Conservancy*, Fall 2003, 35–41.

Stratton-Porter, Gene. "All Together, Heave!" *Izaak Walton League Monthly* 1, no. 4 (1922): 149, 173.

———. *At the Foot of the Rainbow*. Indianapolis: Bobbs-Merrill, 1907.

———. *Birds of the Bible*. Cincinnati: Jennings and Graham, 1909.

———. *Freckles*. New York: Doubleday, 1904.

———. *Friends in Feathers* (expanded version of *What I Have Done with Birds*). New York: Doubleday, 1917.

———. *A Girl of the Limberlost*. New York: Doubleday, 1909

———. *The Harvester*. New York: Doubleday, 1911.

———. "Hidden Treasures . . . An Intimate Study of the Moths of Limberlost Swamp." *Country Life in America* 22, no. 4 (1912): 29–36.

———. *Homing with the Birds*. New York: Doubleday, 1919.

———. *Laddie*. New York: Doubleday, 1913.

———. *Moths of the Limberlost*. New York: Doubleday, 1912.

———. *Music of the Wild*. New York: Eaton & Mains, 1910.

———. "My Work and My Critics." *Bookman* (London) 49 (1916): 147–55.

———. *The Song of the Cardinal*. Indianapolis: Bobbs-Merrill, 1903.

———. *Tales You Won't Believe*. Garden City, NY: Doubleday, 1925.

———. *What I Have Done with Birds*. Indianapolis: Bobbs-Merrill, 1907.

Streshinsky, Shirley. *Audubon: Life and Art in the American Wilderness*. New York: Villard Books, 1993.

Sullivan, Rosemary. *Theodore Roethke: The Garden Master*. Seattle: University of Washington Press, 1975.

Terborgh, John. "Why We Must Bring Back the Wolf." *New York Review of Books*, July 15, 2010, 35–37.

Thoreau, Henry David. *The Journal of Henry D. Thoreau: In Fourteen Volumes Bound as Two*. Edited by Bradford Torrey and Francis H. Allen. 1906. Reprint, New York: Dover, 1962.

———. *Walden: A Fully Annotated Edition*. Edited by Jeffrey S. Cramer. New Haven, CT: Yale University Press, 2004.

———. "Walking." In *Nature Writing: The Tradition in English*, edited by John Elder and Robert Finch. New York: Norton, 2002.

Trachtenberg, Alan. *Shades of Hiawatha: Staging Indians, Making Americans, 1880–1930*. New York: Hill and Wang, 2004.

Treuer, David. "How Do You Prove You're an Indian?" *New York Times*, December 21, 2011. http://www.nytimes.com/2011/12/21/opinion/for-indian-tribes-blood-shouldnt-be-everything.html.

———. *Native American Fiction: A User's Manual*. Saint Paul, MN: Graywolf Press, 2006.

Tyler, Ron. "The Publication of *The Viviparous Quadrupeds of North America*." In Boehme, *John James Audubon in the West*, 119–82.

Van Noy, Rick. *Surveying the Interior: Literary Cartographers and the Sense of Place*. Reno: University of Nevada Press, 2003.

Vaughn, Alden. Introduction to *New England's Prospect*, by William Wood. 1635. Reprint, Amherst: University of Massachusetts Press, 1977.

Vizenor, Gerald. *Manifest Manners: Warriors of Survivance*. Hanover, NH: Wesleyan University Press, 1994.

———. *Summer in the Spring: Ojibwe Lyric Poems and Tribal Stories*. Minneapolis: Nodin Press, 1981.

Wallace, Frank N. "Afield with Gene Stratton-Porter." *McCall's* 53 (June 1926): 109.
Weidensaul, Scott. *Living on the Wind: Across the Hemisphere with Migratory Birds*. New York: North Point Press, 1999.
Wessels, Tom. *Forest Forensics: A Field Guide to Reading the Forested Landscape*. Woodstock, VT: Countryman, 2010.
———. *Reading the Forested Landscape: A Natural History of New England*. Woodstock, VT: Countryman, 1997.
Westling, Louise, ed. *The Cambridge Companion to Literature and the Environment*. Cambridge: Cambridge University Press, 2014.
Wheatcroft, John. "Naughty Child: Poet on Top of a Greenhouse." *North Dakota Quarterly* 66, no. 1 (1999): 60–68.
Whitman, Walt. "Starting from Paumanok." *Leaves of Grass*. New York: Doubleday, Page, & Co., 1917.
Williams, Harry. *"The Edge Is What I Have": Theodore Roethke and After*. Lewisburg, PA: Bucknell University Press/Associated University Presses, 1977.
Williams, Ted. "Something's Fishy." *Audubon*, May/June 2004, 26–33.
Wolff, George. *Theodore Roethke*. Boston: Twayne, 1981.
Wood, William. *New England's Prospect*. 1635. Reprint, Amherst: University of Massachusetts Press, 1977.
Wordsworth, William. "Lines Composed a Few Miles Above Tintern Abbey." In *The Poems*, vol. 1, edited by John D. Hayden, 357–62. New York: Penguin, 1990.

About the Author

Christian Knoeller has taught college English in the Midwest since completing his PhD at UC Berkeley in 1993 and is currently associate professor of English at Purdue where he teaches Reading the Midwest: Engaging Place in Literature and Teaching, a graduate seminar examining midwestern literature from an interdisciplinary, ecocritical perspective emphasizing environmental history. He has received the Midwestern Heritage Prize for Literary Criticism and the Gwendolyn Brooks Prize for Poetry from the Society for the Study of Midwestern Literature (SSML). He is also the author of the scholarly monograph *Voicing Ourselves* in the State University of New York (SUNY) Series in Literacy, Culture, and Learning and *Completing the Circle*, a collection of poems awarded the Millennium Prize. In addition, he has served as president of the Society for the Study of Midwestern Literature.

Index